European Federation of Corrosion
Publications
NUMBER 10

A Working Party Report on

Marine Corrosion of Stainless Steels: Chlorination and Microbial Effects

Published for the European Federation of Corrosion by The Institute of Materials

THE INSTITUTE OF MATERIALS
1993

Book Number 546
Published in 1993 by The Institute of Materials
1 Carlton House Terrace, London SW1Y 5DB

© 1993 The Institute of Materials

All rights reserved

British Library Cataloguing in Publication Data
Available on request

Library of Congress Cataloging in Publication Data
Available on application

ISBN 0-901716-33-2

Neither the EFC nor The Institute of Materials are responsible
for any views expressed which are the sole responsibility of the authors

Design and production by
PicA Publishing Services, Drayton, Nr Abingdon, Oxon

Made and printed in Great Britain
by Biddles Ltd, Guildford and King's Lynn

Contents

Series Introduction — viii

Introduction — ix

INTRODUCTORY PAPERS

1. Aspects of Marine Corrosion and Testing for Seawater Applications — 1
    F. P. IJsseling

2. Microbial and Biochemical Factors Affecting the Corrosion Behaviour of Stainless Steels in Seawater — 21
    V. Scotto, M. Beggiato, G. Marcenaro and R. Dellepiane

EXPERIENCE

3. Marine Corrosion Tests on Metal Alloys in Antarctica: Preliminary results — 36
    G. Alabiso, U. Montini, A. Mollica, M. Beggiato, V. Scotto, G. Marcenaro and R. Dellepiane

4. North Sea Experience with the Use of Stainless Steels in Seawater Applications — 48
    R. Johnsen

CHLORINATION

5. Seawater Chlorination — 60
    C. Madec, F. Quentel and R. Riso

6. Experiences with Seawater Chlorination on Copper Alloys and Stainless Steels — 73
    P. Gallagher, A. Nieuwhof and R. J. M. Tausk

7. Corrosion of Stainless Steels caused by Bromine Emission from Chlorinated Seawater 92
   J. W. Oldfield and B. Todd

TESTING

8. Improved Method for Measuring Polarisation Curves of Alloys During Prolonged Times of Exposure 100
   F. P. IJsseling

9. Application of Electrochemical Impedance Spectroscopy to Monitor Seawater Fouling on Stainless Steels and Copper Alloys 108
   D. Féron

10. Effect of Temperature on Inititation, Repassivation and Propagation of Crevice Corrosion of High-Alloyed Stainless Steels in Natural Seawater 114
    S. Valen, P. O. Gartland and U. Steinsmo

11. An Intelligent Probe for *in situ* Assessment of Susceptibility of Hydrogen Induced Cracking of Steel for Offshore Platform Joints 128
    W. Wei, D. Peng, F. Chao, L. Zheng and D. Yuan-long

12. Aspects of Testing Stainless Steels for Seawater Applications 134
    P. O. Gartland

13. Biofilm Monitoring in Seawater 149
    A. Mollica, E. Traverso and G. Ventura

MECHANISM

14. Identification of Sulphated Green Rust 2 Compound Produced as a Result of Microbially Induced Corrosion of Steel Sheet Piles in a Harbour 162
    J.-M. R. Génin, A. A. Olowe, B. Resiak, N. D. Benbouzid-Rollet, M. Confente and D. Prieur

15. The Role of Green Rust Compounds in Aqueous Corrosion of Iron in Aggressive Media Close to a Marine Environment 167
    Ph. Refait, J.-M. R. Génin and A. A. Olowe

Contents

PROTECTION

16. The Cathodic Protection of Steel for Offshore Platforms in Polluted Seawater   190
D. Yuan-long, F. Chao, L. Zheng and W. Wei

17. Thermal Sprayed Aluminium Coatings in Seawater with and without Cathodic Protection   195
P. O. Gartland and T. G. Eggen

European Federation of Corrosion Publications
Series Introduction

The EFC, incorporated in Belgium, was founded in 1955 with the purpose of promoting European co-operation in the fields of research into corrosion and corrosion prevention.

Membership is based upon participation by corrosion societies and committees in technical Working Parties. Member societies appoint delegates to Working Parties, whose membership is expanded by personal corresponding membership.

The activities of the Working Parties cover corrosion topics associated with inhibition, education, reinforcement in concrete, microbial effects, hot gases and combustion products, environment sensitive fracture, marine environments, surface science, physico-chemical methods of measurement, the nuclear industry, computer based information systems and corrosion in the oil and gas industry. Working Parties on other topics are established as required.

The Working Parties function in various ways, e.g. by preparing reports, organising symposia, conducting intensive courses and producing instructional material, including films. The activities of the Working Parties are co-ordinated, through a Science and Technology Advisory Committee, by the Scientific Secretary.

The administration of the EFC is handled by three Secretariats: DECHEMA e. V. in Germany, the Société de Chimie Industrielle in France, and The Institute of Materials in the United Kingdom. These three Secretariats meet at the Board of Administrators of the EFC. There is an annual General Assembly at which delegates from all member societies meet to determine and approve EFC policy. News of EFC activities, forthcoming conferences, courses etc. is published in a range of accredited corrosion and certain other journals throughout Europe. More detailed descriptions of activities are given in a Newsletter prepared by the Scientific Secretary.

The output of the EFC takes various forms. Papers on particular topics, for example, reviews or results of experimental work, may be published in scientific and technical journals in one or more countries in Europe. Conference proceedings are often published by the organisation responsible for the conference.

In 1987 the, then, Institute of Metals was appointed as the official EFC publisher. Although the arrangement is non-exclusive and other routes for publication are still available, it is expected that the Working Parties of the EFC will use The Institute of Materials for publication of reports, proceedings etc. wherever possible.

The name of The Institute of Metals was changed to The Institute of Materials with effect from 1 January 1992.

A. D. Mercer
EFC Scientific Secretary,
The Institute of Materials, London, UK

Series Introduction

EFC Secretariats are located at:

Dr J A Catterall
European Federation of Corrosion, The Institute of Materials, 1 Carlton House Terrace, London, SW1Y 5DB, UK

Mr R Mas
Fédération Européene de la Corrosion, Société de Chimie Industrielle, 28 rue Saint-Dominique, F-75007 Paris, FRANCE

Professor Dr G Kreysa
Europäische Föderation Korrosion, DECHEMA e. V., Theodor-Heuss-Allee 25, P.O.B. 150104, D-6000 Frankfurt M 15, GERMANY

Introduction

This volume contains the papers which were presented at two recent symposia:

International Workshop on Stainless Steels and Chlorination in Seawater, Brest, 29–30 May, 1991 (EFC Event no. 186), organised by IFREMER Centre de Brest with support from the EFC Working Party on Marine Corrosion.

International Symposium on Marine and Microbial Corrosion, Stockholm, 30 September–2 October, 1991 (EFC Event no. 184), organised by the Swedish Corrosion Institute with support from the EFC Working Parties on Microbial Corrosion and Marine Corrosion.

Many topics of current interest were covered, including papers on microbial corrosion, testing and monitoring, mechanisms, chlorination of seawater, the application of stainless steels in seawater environment and corrosion protection.

Some of the papers presented at the conference were omitted from this publication since they had been published elsewhere.

The majority of the papers in the present volume represent the experience of European authors, although two papers originate from the People's Republic of China. A broad classification of topics had been made, thus Introductory Papers are followed by Experience Papers and then by papers on the specific subjects of Chlorination, Testing Methods (including monitoring), Mechanisms and Protection. The classification is not exclusive since many papers could be associated with more than one topic.

Both conferences demonstrated the necessity to attack the marine corrosion problem in an interdisciplinary way.

F. P. IJsseling
Chairman, Marine Corrosion Working Party

D. Thierry
Chairman, Microbial Corrosion Working Party

INTRODUCTORY PAPERS

Aspects of Marine Corrosion and Testing for Seawater Applications

F. P. IJSSELING

Corrosion Laboratory, Royal Netherlands Naval College (Harssens), c/o Marinepostkamer, PB 10 000, 1780 CA, Den Helder, The Netherlands

Abstract

A brief review is given of the main factors which make seawater such a corrosive fluid. Some modern developments in marine corrosion control and the peculiarities of marine corrosion testing are discussed.

1. Introduction

The corrosive nature of seawater has already been widely documented [1–7]. The main factors which make seawater such a corrosive fluid are now discussed briefly. These factors are divided in two groups: (bio)chemical (i.e. oxygen, carbonate, salts, organic compounds, biochemical activity and pollutants) and physical (i.e. temperature, flow velocity, potential, pressure and light).

The methods of corrosion control are reviewed with emphasis on those methods which find a wide application in the marine environment.

Finally, corrosion testing in general and the peculiarities of marine corrosion testing, in particular the use of natural seawater, stored or recirculated seawater and synthetic solutions for testing purposes is discussed.

2. Corrosion in Seawater

Corrosion in seawater is due to the interaction between the material and the environment, caused by electrochemical reactions which occur at the metal surface. It is well known that corrosion is not a specific materials property such as density, but depends on a large number of variables. A number of these are connected with the alloy, e.g. overall chemical composition, microstructure and surface state. Others are related to the environment, such as type and concentration of redox system(s) which may provoke the corrosion reaction, the presence of compounds which form stable complexes with the dissolved metal ions, etc.

Finally there are a number of physical factors which—depending on the particular system—may exert a considerable influence. Examples of these are potential, temperature, stress and flow velocity.

Generally seaw*ater from the open seas* can be considered as a dynamic aqueous system, containing dissolved salts, gases and organic compounds, undissolved material and living organisms. The dissolved inorganic material comprises almost all known elements, sometimes found in several ionic and molecular forms (Table 1).

The major constituents, as shown in the table, account for over 99.85% of the total dissolved salts. The variations in the concentrations of these species with location and depth are generally rather small. Of the minor components dissolved O_2 and CO_2 are particularly important. Dissolved CO_2 is part of the well-known carbonate/bicarbonate equilibrium reactions (Fig. 1),

Table 1 Concentrations of major chemical compounds in seawater of 35 ‰ salinity [4]

Ion or molecule	concentration (g kg^{-1})
Na^+	10.77
K^+	0.399
Mg^{2+}	1.290
Ca^{2+}	0.412
Sr^{2+}	0.008
Cl^-	19.354
Br^-	0.067
F^-	0.0013
HCO_3^-	0.140
SO_4^{2-}	2.712
$B(OH)_3$	0.0257

$$CO_2(atm) \rightleftarrows CO_2(sol)$$
$$CO_2(sol) + H_2O \rightleftarrows H_2CO_3 \quad \text{(comparatively slow)}$$
$$H_2CO_3 \rightleftarrows H^+ + HCO_3^- \ (K_1) \quad \text{(very fast)}$$
$$HCO_3^- \rightleftarrows H^+ + CO_3^{2-} \ (K_2) \quad \text{(very fast)}$$
$$HCO_3^- + H_2O \rightleftarrows H_2CO_3 + OH^- \quad \text{(fast)}$$
$$Ca^{2+} + CO_3^{2-} \rightleftarrows CaCO_3 \quad \text{(comparatively slow)}$$

Fig. 1 Equilibrium reactions for the system carbon dioxide and water [7].

which form the main basis for the buffering capacity of seawater and its relatively high pH. The pH may vary somewhat with depth and with temperature (Fig. 2). Seawater can be considered to be a bicarbonate solution in equilibrium with the atmospheric carbon dioxide at the surface, and often supersaturated with calcium carbonate, except at great depths and in cold polar regions. The oxygen concentration depends on factors such as temperature and salinity (Fig. 3). Biochemical processes are also involved and the concentrations of dissolved carbon dioxide, oxygen and hydrogen ions are closely coupled through the processes of photosynthesis and biochemical oxidation (respiration):

$$CH_2O + O_2 \rightarrow CO_2 + H_2O$$

biochemical oxidation $\rightarrow$
$\leftarrow$ photosynthesis

Upon the decomposition of carbohydrates, dissolved oxygen is consumed and carbon dioxide is produced, which in turn lowers the pH and decreases the saturation state with respect to carbonates. So in certain areas, e.g. parts of the North Sea, the surface concentration may vary seasonally (Fig. 4). Due to biochemical oxidation lower concentrations may be found at intermediate depths, while at greater depths relative constant concentrations are found due

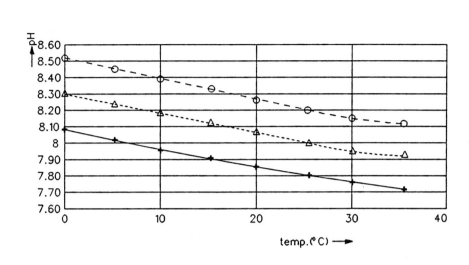

Fig. 2 Variation of pH with temperature as calculated for pH 7.8–8.0 and 8.2 at 25 °C [4].

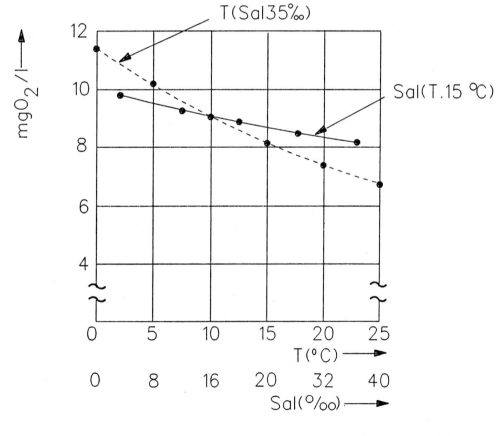

Fig. 3 Variation of oxygen concentration with salinity and temperature [4].

to massive seawater flow. So at a number of locations in the oceans due to this interrelationship a correlation between pH and O_2-content as a function of depth is observed as shown in Fig.5.

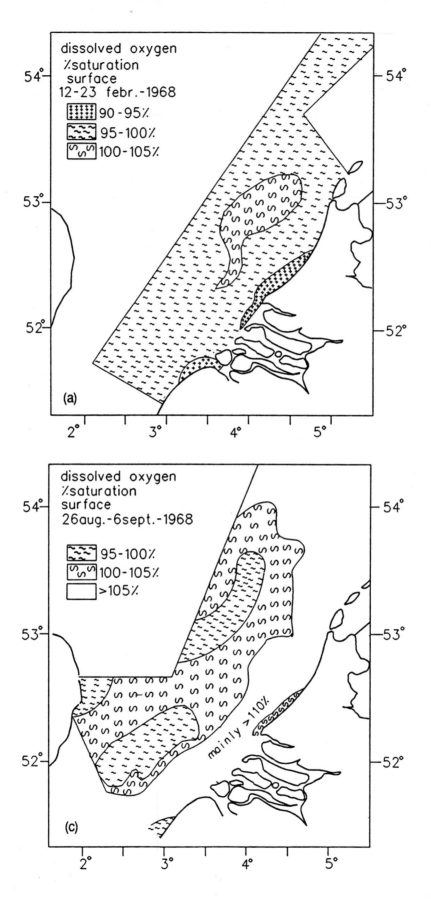

Fig. 4 Seasonal effects on dissolved oxygen concentration as measured during four periods in 1968

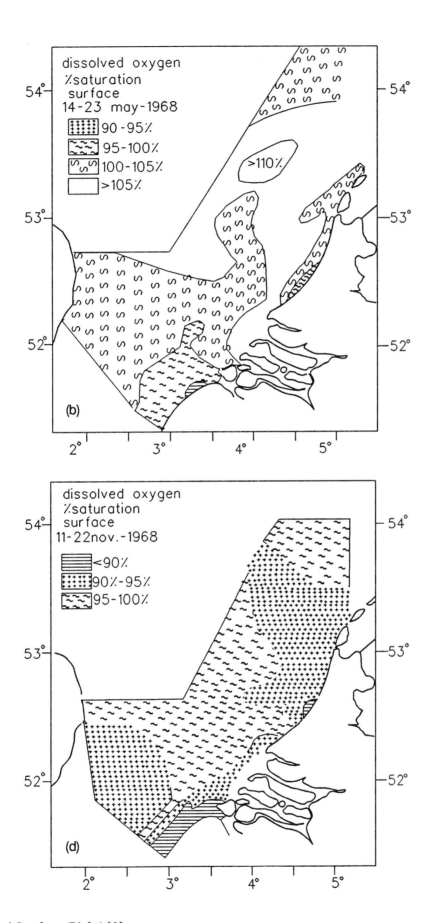

(North Sea / Southern Bight) [8].

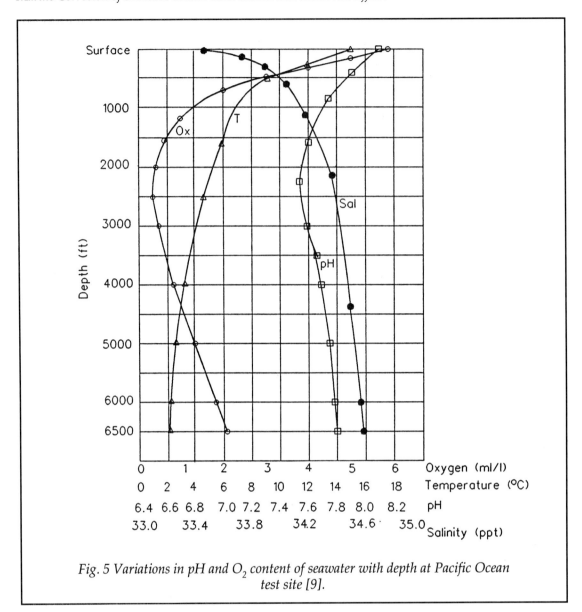

Fig. 5 Variations in pH and O_2 content of seawater with depth at Pacific Ocean test site [9].

At low O_2 concentration anaerobic bacteria may convert sulphates into sulphides.

Listed below are those chemically and biologically oriented factors which are important in relation to the corrosivity of seawater:

1. The oxygen content: in normal seawater oxygen is generally the chemical compound which is being reduced in the cathodic reaction, the hydrogen ion concentration being relatively low. So, basically, the corrosion rate should be linearly dependent on the oxygen concentration, as has been observed in a number of cases. However, under practical conditions the formation of layers or films often interferes with this simple concept. Other exceptions to this general trend are: (a) alloys which tend to passivate require oxygen to function in the passive state with low corrosion rate, and, (b) alloys which have rather active corrosion potentials may corrode with a cathodic reaction involving reduction of hydrogen ions and/or water itself.

2. Dissolved carbon dioxide which, via the dissociation in bicarbonate and carbonate ions

is mainly responsible for the buffering of surface seawater at about pH 8. The weak basic nature of seawater and its buffering capacity are frequently instrumental in the formation of corrosion product and other types of layers.

3. The high dissolved salt content (salinity). This has two actions: (a) the main part of the dissolved species are chloride ions, which generally promote metal dissolution by complexation; moreover, chloride ions, being detrimental with regard to passive films, also promote local breakdown of such films, leading to forms of local attack, such as pitting, crevice corrosion and intergranular corrosion; (b) a high electrical conductivity, leading to larger currents for a given potential difference. Examples of this are larger galvanic effects and larger surfaces which pass cathodic currents in the event of local corrosion attack.

4. The presence of calcium and magnesium ions which, in combination with the relatively high seawater pH, are often involved in the formation of more or less protective layers. This effect is often amplified by an increase of the pH at the metal/solution interface due to the cathodic reaction, mass transfer conditions also being involved in determining the local pH and a possible gradient of pH (Fig. 6).

5. The presence of organic compounds, of which there are thousands and which generally also act via their complex forming properties with metal ions.

6. Biological activity, which has many possibilities for provoking and accelerating corrosion reactions, such as:

(i) the production of corrosive metabolic products (such as acids and hydrogen sulphide),
(ii) the formation of discontinuous deposits on the surface, resulting in differential aeration and concentration cells,
(iii) disruption of natural and other synthetic films,
(iv) breakdown of corrosion inhibitors and coatings,
(v) depolarisation of electrochemical reactions, etc.

The above-mentioned effects may lead to initiation of corrosion, changes in the mode of corrosion (i.e. from uniform to local attack), and to changes in corrosion rate [11,12].

7. Closely related to this is the possible presence of sulphates and hydrogen sulphide, often linked to the presence of organic pollutants and low oxygen content. The former, although generally non-corrosive, may act indirectly due to transformation to sulphides by the action

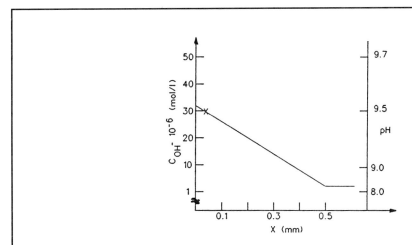

Fig. 6 Concentration profile of OH⁻ at the surface, assuming Nernst diffusion layer and natural convection (pH = 9.5 at distance 25 µm) [10].

of sulphate-reducing bacteria. In particular, with copper alloys this may give rise to very unfavourable conditions [13,14].

The presence of heavy metal ions, which may reach significant concentrations in, for example, harbour mud or areas which are not regularly flushed down, etc. In particular, the contact of aluminium alloys with dissolved copper ions is a well known example of a potential corrosion hazard.

Physically oriented factors include:

1. Temperature: since corrosion reactions in most cases are mass transfer and not kinetically controlled the temperature effect is relatively limited. Moreover, increasing temperature will lead to decreasing oxygen solubility, which will tend to counteract any temperature influence. Otherwise temperature may have a large influence on protectiveness, of, e.g. corrosion product layers (Fig. 7) and on the role of biological processes.

2. Mass transfer and flow: all action which promotes transport of oxygen to the metal surface will enhance corrosion effects and in the case of actively corroding alloys the result will be detrimental. On the other hand, as oxygen can promote passivity, the effect will generally be advantageous in the case of passive alloys. Flow conditions often interfere with corrosion reactions because of their possible interaction with film formation. Unevenly distributed films, or those providing incomplete surface coverage, may give rise to the formation of macro or micro corrosion cells.

3. Potential: as corrosion reactions in seawater are of an electrochemical nature potential will be expected to have a large influence on the corrosion rate. However, the effect tends not to be pronounced in practice since most seawater corrosion systems are mass transfer controlled and metal surfaces are often covered with layers of corrosion products.

4. Pressure: it has been found that the corrosion of certain alloys may depend on depth, not

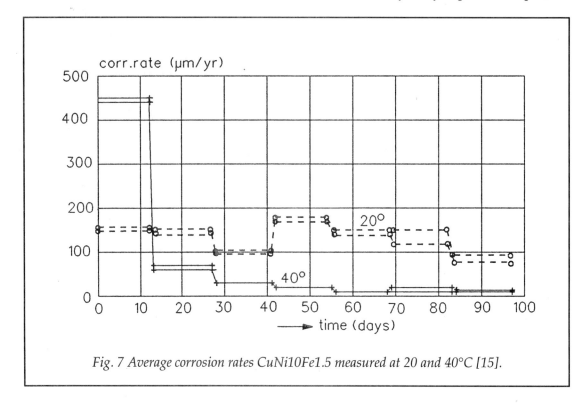

Fig. 7 Average corrosion rates CuNi10Fe1.5 measured at 20 and 40°C [15].

only because of the different conditions regarding pH, oxygen content and temperature, but also because of different compositions and morphologies of the corrosion products [16, 17].

5. Light and illumination may have a strong effect on photochemical reactions such as photosynthesis. This is also related to the levels of ecosystems present.

Many of the above-mentioned factors are interrelated, and some relevant combinations should be noted, e.g. oxygen solubility and temperature: the oxygen effect in many practical systems tends to predominate above the temperature effect if only these factors are taken into account.

Generally the corrosion of metals in seawater is strongly affected by the formation of films on the metal surface. These films may be passive films, corrosion product layers, calcareous layers, bacterial films and macrofouling layers or combinations thereof.

Some typical examples include:

- biological slime on stainless steel,
- effect of hydrogen sulphide on copper alloys covered with corrosion product,
- macrofouling on mild steel,
- calcareous layers on cathodically protected surfaces.

From these examples it is obvious that the effect of the layers is very dependent on both the nature of the layer itself and the underlying metal. The presence of a layer means that the metal surface is in contact with a micro environment which may be quite different from that on top of the layer. The nature of the layer, i.e. its composition, morphology, porosity and adherence to the substrate will play a part. Depending on the conditions the corrosion rate may increase or decrease, while the nature of the corrosion process may also change, e.g. from uniform to local corrosion. The presence of layers may also induce the development of macro corrosion cells when the layer is not distributed evenly over the whole exposed surface, or of micro cells when local imperfections or uncovered areas are present. So the effect of the layers may be very variable. For this reason, the ability to control and monitor layer formation is often essential for corrosion control. For example, the impact of calcareous layers on the required cathodic protection current has been studied extensively with respect to the effects of external factors, such as temperature and flow velocity [18, 19] on their morphology and protective nature.

Although the composition of seawater does not vary greatly worldwide, there are a few minor compounds which exert an overriding influence on the corrosion process. The concentrations of these minor compounds depend on location, and vary—particularly in coastal and/or polluted areas—with the season and with depth. The influence of these minor components on the corrosion process may be direct or indirect via the formation or breakdown of some kind of layer. So to estimate the corrosivity of the seawater for a given application it is essential to determine the concentration of dissolved oxygen, carbon dioxide, chloride, pH and the possible presence of pollutants from domestic, industrial or agricultural sources. The possibility of fluctuations due to seasonal effects, flow conditions, etc. should also be considered. Apart from the above mentioned variables, when selecting materials for a newly erected plant or installation, consideration should be given to possible changes in the future due to forthcoming developments, e.g. the construction of other industrial sites.

Due to the many variables involved, corrosion may manifest itself in several forms, which may be broadly separated into uniform and non-uniform (local) types of corrosion. In the latter case there is always some kind of non-homogeneity which exerts a predominating effect. Such non-homogeneities may be present in the metal phase itself, in the solution phase or in other areas of relevance. Examples are precipitates in the alloy microstructure, differential aeration, locally different stress or flow conditions, potential differences, etc. These are the underlying causes for the corrosion process to occur preferentially at one site or zone. It is well-known that these local corrosion processes are much more troublesome in practice than their uniform counterpart since they frequently lead to dangerous situations. Often the susceptibility to localised corrosion types can be connected with specific alloys or alloy groups, i.e. stainless steels to pitting, crevice corrosion and intergranular corrosion, copper alloys to erosion corrosion, etc. However, this is a very general statement and within groups of materials there may be large differences between alloys, depending to a large extent on their composition and microstructure.

3. Corrosion Control

Corrosion is to be regarded as an enemy, of which one must be continuously aware. The following general routes are open to combat corrosion and to control its detrimental effects:

(i) materials selection,
(ii) adhering to sound engineering design rules,
(iii) modifying the environment, for instance by removing oxygen, modifying the pH, etc.,
(iv) the addition of inhibiting or scale-removing or biologically active compounds,
(v) the application of coatings,
(vi) the application of cathodic (or anodic) protection.

In practice combinations of these possibilities are often applied.

Materials selection is, of course, closely related to materials development. In the development sphere over the last decades we have seen some tremendous achievements in improved fabrication methods and product control of various materials, mainly due to improved insight in the interrelationship between overall chemical composition, microstructural aspects and materials properties and behaviour. This statement holds for alloys as well as composite materials. These improvements have lead to the concept of materials design, i.e. the design and fabrication of materials with specific properties for specific applications.

Mild steel, which is still the most widely used metal for constructional purposes, is available in thick plate form, with excellent cross-sectional homogeneity and good through-thickness properties. There has also been the development of higher strength steels, often containing a small amount of alloying elements and needing some thermomechanical treatment. All these steel types corrode readily in seawater and some kind of protection is required, unless the thickness is sufficient to ensure mechanical strength during the envisaged technical use.

Great developments have been made in the stainless steel sector, where modern technology has produced a range of alloys with outstanding corrosion resistance, mechanical strength and weldability [20]. For specific applications some of these stainless steel types are used to replace older alloys. In seawater transport systems copper alloys, particularly of the copper–nickel type, have been used traditionally. However, the application of stainless steels in this area is increasing, due to the significantly higher flow velocities which can be tolerated. The main incentive is a substantial decrease in weight, with regard to the system itself and its water content. In the case of off-shore platforms weight above water is significantly related to cost and in the end a large amount of money can be saved by weight reductions. However, as the stainless steels are far more prone to marine growth than copper alloys, additional measures have to be taken, e.g. the dosing of biocides [21–24].

Copper alloys have also seen important developments: particularly in the range of castable high strength copper nickel to replace nickel aluminium bronzes, which in some high-tech application shows some disappointing reliability due to selective corrosion attack [25, 26]. Of course, there have also been new developments in other alloy types, such as nickel-, aluminium- and titanium alloys, which space does not allow to be discussed in detail.

There are also many new possibilities arising from composite materials with a resin matrix [27, 28]. These are resistant to the normal forms of corrosion. However, they have other properties which are more or less detrimental or still not fully understood. Their relatively high fabrication cost and a lack of standardisation still prevent their application on a much larger scale, though there is an ever increasing stream of applications, especially in military applications.

Another notable method of corrosion control is the use of a number of rather simple design rules, which prevent conditions liable to provoke corrosion. However important this may be in practice, the topic will not be developed here as it is outside the scope of this symposium.

For the control of marine corrosion modification of the environment is not generally applicable, as this is limited to closed systems, like seawater distillation equipment, etc. Generally, the same holds for the addition of inhibiting and/or scale removing substances, although in a limited number of cases such actions are being applied successfully. On the other hand the addition of biologically active components is being used on a much larger scale, as shown by the number of papers dealing with this subject in this volume. Nevertheless, these are special applications, e.g. in seawater storage and transport systems, which still can be regarded as semi-closed systems.

The application of coatings remains a very important method of corrosion protection. Coatings vary between metallic (in the form of thin layers as well as linings), and non-metallic such as synthetic resins, rubbers, tar, concrete, paint, ceramics, mortars, etc., or combinations of these.

Some examples of metallic coatings may be given here: copper–nickel is used as sheathing for offshore platform piling [29], while its potential use as cladding for ship plates is under investigation [30]. Flame sprayed aluminium (FSA) has received much attention in the last decade. It is being used extensively for protection against atmospheric corrosion (inside and outside of ships) by the U.S. Navy [31]. An example of subsea use is in corrosion control of the tension leg components and production risers of Conoco's tension leg platform at the Hutton field in the North Sea [32, 33]. The FSA coating was sealed by a vinyl- or a silicon-type sealer and was designed to provide a 20-year life in the aggressive North Sea conditions. On inspection after three years service it was found that the FSA coating performed satisfactorily, the rate of consumption not being excessive and its design life expected to be reached. There was some concern about the appearance of blisters, but only 5% of these had penetrated to the underlying steel substrate, which nevertheless showed no signs of pitting or increased general attack. The blistering was found to be due to a combination of factors, the main cause being inadequate sealing of the aluminium by the vinyl sealer [34]. The good results are due to the FSA system performing both as a tough barrier type coating and as a sacrificial anode. The application of cathodic protection to FSA coated steel has been investigated, the outcome being that a very limited current is required to maintain the potential at the desired protection level [35].

In the field of organic coatings much effort has been dedicated to obtaining a better insight into the fundamental aspects of adherence to the substrate, the required surface preparation, the application technology and its control, the possible causes of degradation, and last but not least, the impact on the environment. Also the compatibility with cathodic protection is a topic for continuing research. Much time has been devoted not only to the development of newly formulated coatings for specified application areas, but also to the technology, requirements and allowed tolerances of surface preparation, including protection of surfaces between surface preparation and application of the coating.

Significant improvements have been made in the technique of cathodic protection, particularly in the translation of the theoretical principle of the method into practical rules and codes of practice [36–41]. With regard to the actual equipment many improvements have been made, both in the field of sacrificial anodes and impressed current systems, and in the methods for control and monitoring of the effectiveness of cathodic protection.

3.1 Trends for the future

Generally it is expected that mechanical strength and useful life will increase, along with a decrease in willingness to spend, time and money on maintenance. Moreover, the impact of environmental restrictions will continue to grow. This may be translated into an increased application of the concept of materials design, increased quality and reliability requirements, increased possibilities for repair and recycling and, last but not least, an ever increasing influence of limitations, set by the environmental requirements, on the aspects of production, application, operation and recycling of materials. This all points to a continuing need for research and development in the field of materials, materials application and materials protection.

4. Marine Corrosion Testing

All improvements in materials and protection methods require testing to begin with, and special attention will be given to this important topic. Testing can be performed for several different purposes, i.e. to collect data on corrosion behaviour and rates or on corrosion mechanisms. There are standardised tests for quality control purposes and ranking tests for comparing several alloys. Another possible way to classify test methods is into laboratory tests (often accelerated) and *in situ* field or in-service tests.

The ultimate choice of a test method—or test methods—depends greatly on the purpose of the test. Generally, accelerated laboratory tests may suffice for a first ranking or screening. However, to be able to say something about the actual service performance, in-service or field tests of sufficiently long duration are essential. Such testing aimed at obtaining data on the corrosion performance of alloys in practical situations is a complicated problem due to the many variables. This is certainly the case for marine applications because of the overriding influence of corrosion products and other layers on the corrosion behaviour.

The choice of the test environment is very important in this area. As we have seen, there can be a large deviation between the corrosivity of seawater at different locations, in particular in coastal areas, where the water may be brackish and/or polluted to quite different degrees. It follows that it is imperative to test in real seawater to obtain 'absolute' corrosion rate values for a given application. Moreover, it is necessary to take local conditions into account since, as we have seen, the influences which affect the corrosion reactions are those that are the most dependent on location, season and a variety of other factors [42].

Apart from using the naturally occurring seawater for testing, it is also necessary to be aware of the peculiarities of seawater. For instance, it is well documented that the use of recirculated or previously stored seawater may also give rise to different results, depending on the alloy system (Figs. 8 and 9). This may be due to changes in the seawater corrosivity itself, brought on by changes in pH and/or the biological activity, or the accumulation of corrosion products. All these possibilities may influence the formation of layers in qualitative and quantitative ways, leading to test results which are unrepresentative. Moreover, when testing under local conditions one must be sure that these will not change due to the technical set-up. A well-known example is an electricity generating plant, which uses seawater for cooling purposes. Although before the construction the seawater to be used was examined and found to be free from pollution, the plant itself was built further inland with a 200 m duct providing the connection with the sea. At the end of this channel, near the plant, a basin was constructed as an emergency precaution. After a few months of operation severe corrosion problems were encountered. It was found that the seawater ultimately supplied to the condensers was heavily polluted with sulphides due to the decay of organic material which had grown in the duct and

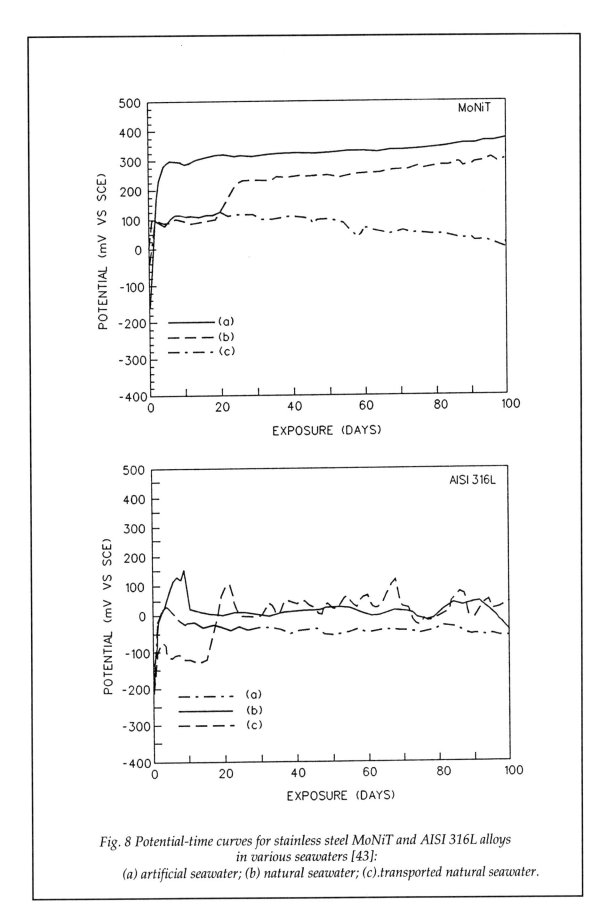

*Fig. 8 Potential-time curves for stainless steel MoNiT and AISI 316L alloys in various seawaters [43]:
(a) artificial seawater; (b) natural seawater; (c).transported natural seawater.*

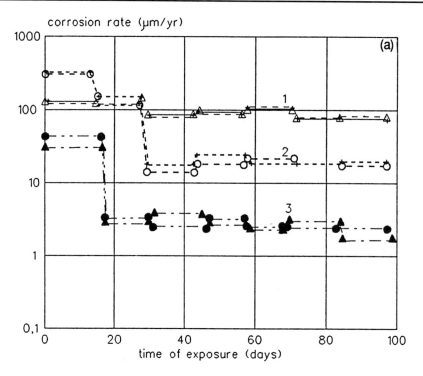
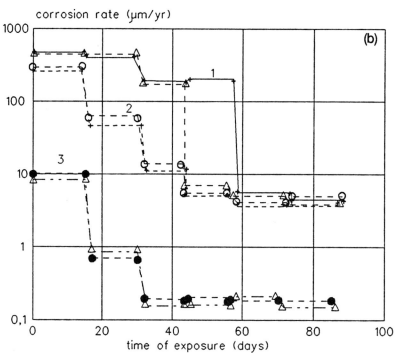

Fig. 9(a) Corrosion rate as a function of time for Al-brass exposed to natural and synthetic seawater at 20°C [44]:
1. nat. seawater/cont. refreshment/100× per day;
2. nat. seawater/cont. refreshment/1× per day;
3. synth. seawater/batch refreshment/1× per day.
(b) As (a), temperature 40°C.

the basin. In the end, quite expensive measures had to be taken to repair the damage and to prevent the sulphide pollution.

Apart from naturally existing seawater, there are also some synthetic mixtures for simulating seawater. The usual compositions contain only the major inorganic components, buffered by the addition of the appropriate amount of carbonates. These mixtures obviously lack the organic and biological compounds which may exert a strong influence on corrosion processes. For this reason such synthetic mixtures should never be used for measuring corrosion rates for practical applications. A few examples are shown in Figs. 9 and 10.

Generally, synthetic simulations and transported and/or recirculated seawater are only suitable for accelerated laboratory tests intended for ranking and screening purposes. Of course, this is a rather general statement. For instance, conditionally stored and/or recirculated seawater might be used for determining seawater corrosion rates. Whether this will be successful depends on the precautions which can be taken to keep the corrosion product concentration low and the corrosivity of the seawater, including its biological activity, at the same level as encountered in practice [46]. Whether this is feasible also depends very much on the actual alloy system to be tested. Another field in which synthetic and stored or treated seawater might be useful is in mechanistic investigations, when attempts are made to isolate the effect of one or more variables which are thought to influence the corrosion reaction.

The remaining part of this chapter covers actual corrosion testing, and in particular some local types of corrosion, i.e. galvanic corrosion, crevice corrosion and pitting.

Galvanic corrosion is defined as the increase in corrosion rate of an alloy due to direct metallic coupling to a more noble metal, both metals being in contact with the same electrolyte. The theoretical understanding of the problem has been much improved, and in particular the ability to predict the intensity of galvanic corrosion by means of mathematical models based on, or related to, the finite element method [47–54]. This has gone hand in hand with the improvement in theoretical insight into cathodic protection—a closely related problem. So, taking into account both geometrical effects and the polarisation behaviour of the alloys constituting the galvanic couple, it is feasible now to give a far more accurate estimate of galvanic effects than was hitherto possible on the basis of the difference in free corrosion potentials only. In practical testing, the use of zero resistance ammeters and potentiostats has made it far simpler to estimate galvanic corrosion rates and to identify probable problem areas [55]. By careful modelling and testing under actual seawater conditions, practical situations can be simulated [56]. Also the measurement of polarisation curves preferentially during long times of exposure may add substantially to our knowledge [57–59]. Recently much work has been devoted to the compatibility of copper alloys with stainless steels, mainly under conditions relevant to seawater systems [60, 61].

Crevice corrosion is a local corrosion type, generally related to the application of passivated alloys: a mathematical model was proposed by Oldfield and Sutton [62]. The crevice corrosion process was divided into four steps:

(i) depletion of oxygen in the crevice,
(ii) decrease of pH and increase of Cl^- concentration in the crevice solution,
(iii) permanent breakdown of passive film inside the crevice, and
(iv) crevice corrosion propagation.

The first three steps constitute the initiation phase. A large number of parameters influence the initiation process, the most important being the alloy composition (i.e. passive current density), crevice geometry, potential, composition of bulk solution and temperature. The fourth step constitutes the propagation phase, an important factor being the decrease of the critical passivation current density on increasing acid and chloride concentrations. During

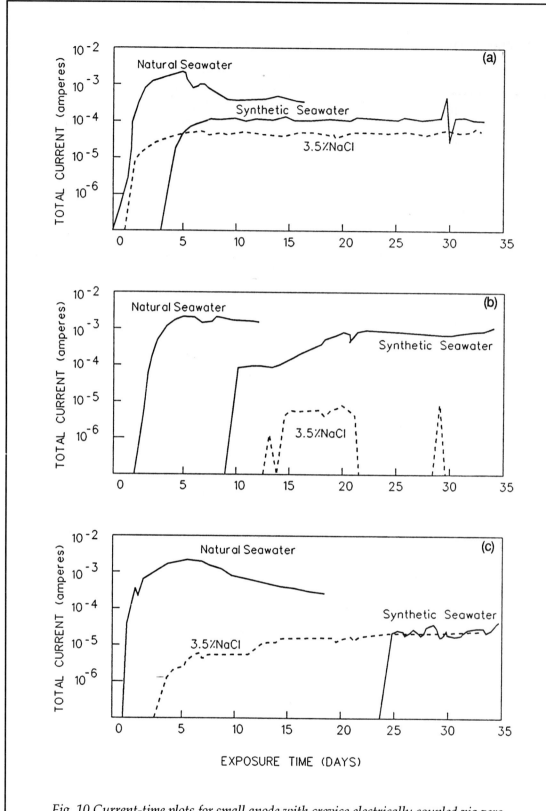

Fig. 10 Current-time plots for small anode with crevice electrically coupled via zero resistance ammeter to a crevice free cathode in natural and synthetic environments [45]: (a) type 316 stainless steel; (b) 18 Cr-2 Mo stainless alloy; (c) alloy 904 L.

Aspects of Marine Corrosion and Testing for Seawater Applications

propagation the reaction rate may be controlled by the anodic or the cathodic reactions or by both. Consequently, factors such as electrochemical reaction rates and solution flow rate become important. The Oldfield–Sutton model is aimed more at the initiation phase. An extension of the crevice corrosion model in which the propagation stage is addressed has been developed by Gartland [63]. In this model the following factors are taken into account:

(i) significance of the outside potential,
(ii) ohmic potential drop along the crevice,
(iii) potential and time dependencies of passive current density,
(iv) initiation mechanism at high outside potentials,
(v) correction of ion mobilities at high ionic strengths, and
(vi) hydrolysis of molybdenum.

Testing for susceptibility to crevice corrosion has also been improved as a result of an improved insight into the factors which may influence the crevice corrosion process. Methods which are being applied include [45, 64–66]:

- the measurement of polarisation curves of crevice-free samples to determine the critical crevice solution, i.e. the chloride concentration and pH at which spontaneous depassivation of the alloy is expected [67];
- long term exposure of samples with crevices under actual conditions [68];
- accelerated tests using samples with crevices often under potentiostatic or potentiodynamic control, to determine the critical crevice corrosion potential and the repassivation potential; such tests can also be performed, by increasing the temperature in steps, to determine the critical crevice temperature [69, 70];
- other tests aimed specifically at determining the initiation time and/or propagation rate under specified conditions (Fig. 9).

Nevertheless, most test methods are mere ranking tests, the correlation with behaviour under practical conditions still having to be defined. A definite problem area in crevice corrosion testing is that of the large influence of crevice geometry and the practical restraints in providing reproducible crevices. For this reason much attention has been paid to the crevice-forming device to be used [71, 72]. One possible way to handle this problem is to use a very large number of small crevices, so increasing the probability that crevice attack will occur [73].

Pitting in many cases is recognised to be a statistically based process, the initial or parent distribution of pit depths being roughly approximated by a normal law, or, after long periods by a log normal law [74, 75]. For practical situations, the maximum pit depth is an important value and often follows an extreme value distribution [76–78]. By application of the appropriate statistical theory the well-known influence of the sample surface on the test results can be taken into account and also estimates can be made of the time to perforation. This concept has been applied among others to the pitting susceptibility of experimental HSLA steels exposed to seawater [79].

In electrochemical testing the critical pitting potential and the critical pitting temperature (CPT) are often used as measures of the pitting susceptibility of an alloy [80, 81]. These are both ranking numbers and must be treated with caution. For instance, the CPT must not be regarded as the maximum temperature at which the alloy can be practically used without danger of pitting attack. Recently, these statistically based concepts have been introduced in the determination of the pitting potential [82]. To take the influence of the sample surface into account an 'elementary pitting probability' (EPP) was defined, characterising the intrinsic behaviour of the alloy without respect to the sample area. The EPP was found to be an

17

exponential function of potential. However, the obtained results are strongly dependent on the applied scanning rate, which itself is related to the alloy composition and probably to the chloride concentration.

An experimental problem which often had a detrimental effect on the accuracy of pitting potential determination was crevice corrosion occurring at the sample edges. To solve this problem some special constructions of the sample holder have been developed [83–86].

These aspects of corrosion testing—in the general sense as well as those dedicated specifically to marine conditions—are necessarily incomplete and give only selective information. In general, it is clear that there has been—and still is—a tremendous improvement in available equipment and techniques. Examples include computer controlled potentiostats and data loggers, including a range of software to process the results obtained. A technique which has emerged as very promising is that of electrochemical impedance spectroscopy (EIS), which, for example, has been applied intensively to organic and inorganic coatings to obtain more insight into the mechanism of protection and degradation.

The same comments made with regard to testing also apply to monitoring. Combinations of sophisticated techniques are also being used to monitor systems prone to corrosion attack in the marine field. The development of sophisticated sensors and equipment for data generation and manipulation have proved very effective for this kind of application. And we are certainly not at the end of this development.

5. Concluding Remarks

Marine corrosion is still a large problem despite the many technological developments in its combating and control. Many aspects still require a deeper understanding. These areas are mainly to be found where the formation of layers takes place at the metal–solution interface.

As layer formation may be influenced by mechanical, chemical, electrochemical, biological and metallurgical factors, a truly interdisciplinary approach is required. It is a pity to see such an approach often neglected. For this reason, it is very important for various groups of workers to be brought together, as indeed was the case at the symposia from which the papers in this volume are drawn. Of course, the approach as well as the tools to be used will differ from one group to another. However, to arrive at the desired goal, a real understanding of the possible interference of biological factors with marine corrosion is imperative: we must understand each other's methods and languages to accomplish this.

References

1. L. L. Shreir (ed.), Corrosion, 2nd edn, Vol. 1, ch. 4, London, Newnes-Butterworths, 1976.
2. F. L. Laque, Marine corrosion, ch. 4. New York, Wiley, 1975.
3. K. A. Chandler, Marine and offshore corrosion', ch. 3. London, Butterworths, 1985.
4. J. P. Riley and R. Chester, Introduction to marine chemistry. London, Academic Press, 1971.
5. S. C. Dexter, in 'Metals handbook', 9th edn, Vol. 13, 893. 1987, Metals Park, OH, ASM International.
6. S. J. Dexter and C. Culberson, Mater. Perform., 1980, **20**, (9), 16.
7. W. F. McIlhenny and M. A. Zeitoun, Chem. Eng., 1969, 81; (1969) 251.
8. S. B. Tijssen, Ann. Biologiques, 1969, **26**, **77.**
9. F. M. Reinhart, Geo-Marine Techn., 1965, 9.
10. R. Holthe, dissertation Univ. of Trondheim, October, 1988, Fig. 5.42.
11. M. Walch and R. Mitchell, Naval Research Reviews, 1984, **3**, 13.
12. L. J. Seed, Corrosion Reviews, 1990, 1-2, 1.
13. K. D. Efird and T. S. Lee, Corrosion, 1979, **50**, 79.
14. C. Kato, H. W. Pickering and J. E. Castle, J. Electrochem. Soc., 1984, **131**, 1227.

15. F. P. IJsseling, paper OS-130, Proc. 9th European Congr. on Corrosion, Utrecht, Oct., 1989.
16. A. M. Beccaria, G. Poggi, M. L. Lorenzetti and G. Castello, paper OS-138, Proc. 9th European Congr., Corrosion, Utrecht, 2–6 Oct. ,1989.
17. D. Festi and A. M. Beccaria, Corrosion NACE '89, New Orleans, 17–21 April, paper 292.
18. S-H Lin and S. C. Dexter, Corrosion, 1988, **44**, 615.
19. K. Nisancioglu, Corrosion, 1987, **43**, 100.
20. A. J. Sedriks, Corrosion, 1989, **45**, 510.
21. B. Wallén, Werkst.u.Korr., 1989, **40**, 602.
22. P. O. Gartland and J. M. Drugli, Corrosion NACE '91, Cincinnati, OH, 11–15 March, paper 510.
23. G. Ventura, E. Traverso and A. Mollica, Corrosion, 1989, **45**, 319.
24. R. Gundersen, B. Johansen and P. O. Gartland, Corrosion NACE '89, New Orleans, Louisiana, 17–21 April, paper 108.
25. E. C. Mantle, MER, 1986, July, 19.
26. S. H. Lo, W. M. Gibbon, R. S. Hollingshead and S. Corbin, Materials & Design, 1987, **8**, (1), 30.
27. R. L. Pegg and H. Reyes, Advanced Materials & Processes, 1987, **131**, (3), 35.
28. M. J. Seamark, Trans. Inst. of Marine Engineering, 1990, **102**, (4), 261.
29. D. G. Melton, Corrosion, 1988, **44**, 478.
30. P. Drodten and H. Pircher, Werkst.u.Korr., 1990, **41**, 59.
31. A. R. Parks, Proc. 2nd Nat. Congress on Thermal Spray, Long Beach, California, 1984, 65.
32. W. H. Thomason, Materials Performance, 1985, (3), 20.
33. M. T. Copper and W. H. Thomason, Anti-Corrosion, 1986, (7), 4.
34. T. Rosbrook, W. H. Thomason and J. D. Byrd, Materials Performance, 1989, (9), 34.
35. P. O. Gartland and T. G. Eggen, Corrosion NACE '90, Las Vegas, Nevada, 23–27 April, paper 367.
36. W. Keim, R. Strommem and J. Jelinek, Materials Performance, 1988, (9), 25.
37. S. Aoki, K. Kishimoto and M. Miyasaka, Corrosion, 1988, **44**, 926.
38. D. J. Tighe-Ford, J. N. McGrath and M. P. Wareham, Trans.I. Mar. E., 1988, **100**, 185.
39. R. Strommen, W. Keim, J. Finnegan and P. Mehdizadeh, Materials Performance, 1987, (2), 23.
40. S. B. Lavlani and V.V. Patel, Corrosion, 1990, **46**, 755.
41. K. Nisancioglu, Proc. 2nd Int. Conf. on Cathodic Protection, Theory and Practice, Stratford-upon-Avon, 26–28 June, 1989.
42. General Guidelines for Corrosion Testing of Materials for Marine Applications, European Federation of Corrosion Publications, No. 3, The Institute of Metals, London, 1989.
43. P. Gallagher, R. E. Malpas and E. B. Shone, Br. Corros. J., 1988, **23**, 229.
44. F. P. IJsseling, Proc. 9th European Congress on Corrosion, Utrecht, 1989, 2–6 Oct., paper OS 130.
45. R. M. Kain and T. S. Lee, in Laboratory Corrosion Tests and Standards, STP 866, G.S. Haynes and R. Baboian, ed., ASTM, Philadelphia, 1985, 299.
46. J. E. Castle, A. H. L. Chamberlain, B. Garner, M. Sadegh Parvizi and A. Aladjen, in 'The use of synthetic environments for corrosion testing', ed. P.E. Francis and T.S. Lee, STP 970, ASTM, Philadelphia, PA, 1988, 1 74.
47. J. W. Fu and Siu-Kee Chan, Materials Performance, 1986, (3), 33.
48. S. U. Fangteng and E. A. Charles, Corr. Science, 1988, **28**, 649.
49. E. Bardal, R. Johnsen and P. O. Gartland, Corrosion, 1984, **40**, 628.
50. R. B. Morris, J. Electrochem. Soc., 1990, **137**, 3039.
51. D. J. Astley, Corr. Science, 1983, **23**, 801.
52. D. J. Astley and J. C. Rowlands, Br. Corros. J., 1985, **20**, 90.

53. R. G. Kasper and M. G. April, Corrosion, 1983, **39**,181.
54. J. W. Fu, Corrosion NACE, 1986, Houston, Texas, 17–21 March, paper 44.
55. S. Valen, E. Bardel, T. Rogne and J. M. Drugli, Proc. 11th Scandinavian Corrosion Congress, Stavanger, 1989, paper F-63.
56. Galvanic Corrosion, ASTM STP 978, H.P. HACK, ed. ASTM, Philadelphia, 1988.
57. H. P. Hack and J. R. Scully, Corrosion, 1986, **42**, 79.
58. G. O. Davis, J. Kolts and N. Sridhar, Corrosion, 1986, **42**, 329.
59. G. Prentice, R. A. Holser, V. J. Farozic, R. B. Pond, Jr and K. L. Cramblitt, Corrosion, 1990, **46**, 77.
60. R. J. Ferrara, L. Ferrara, L. E. Taschenberg and P. J. Moran, NACE Corrosion'85, Boston, Massachusetts, paper 211.
61. B. Wallén, Werkst. u. Korr., 1989, **40**, 602.
62. J. W. Oldfield and W. H. Sutton, Br. Corros. J., 1978, 13, (1), 13 and (3), 104.
63. P. O. Gartland, R. Holthe and E. Bardal, Proc. 11th Scand. Corros. Congr., Stavanger, 1989, paper F-64.
64. J. W. Oldfield, Int. Materials Review, 1987, **32**, (3),1.
65. F. P. IJsseling, Br. Corros. J., 1980, **15**, 51.
66. M. Hubbell, C. Price and R. Heidersbach, ibid., 324.
67. M. D. Carpenter, R. Francis, L. M. Philips and J. W. Oldfield, Br. Corros. J., 1986, **21**, 45.
68. M. A. Streicher, Materials Perform., 1983, **22**, (5), 37.
69. P.O. Gartland and S. M. Valen, Corrosion NACE, Cincinnati, Ohio, 11–15 March, paper 511.
70. S. Valen and P. O. Gartland, Proc. Eurocorr' 91, Budapest, 21–25 Oct., 1991.
71. R. J. Brigham, Corrosion, 1981, **37**, 608.
72. P. O. Gartland, Proc. Symp. on Marine and Microbial Corrosion, Stockholm, 30 Sept.–2 Oct., 1991.
73. J. M. Krougman and F.P. IJsseling, in 'Electrochemical Corrosion testing', E. Heitz, J. C. Rowlands and F. Mansfeld eds, Dechema Monographie no. 101, Frankfurt-M., 985, 135.
74. V.Ya. Flaks, Transl. Zaschita Metallov, 1973, **9**, 443.
75. M. Janik-Czachor and M. B. Ives, in Passivity of Metals, ed. R.P. Frankenthal and J. Kruger, The Electrochemical Soc., Princeton, New Jersey, 1978, 369.
76. P. M. Aziz, Corrosion, 1956, **12**, 495.
77. A. K. Sheikh, J. K. Boah and D. A. Hansen, Corrosion, 1990, **46**,190.
78. P. J. Laycock, R. A. Cottis and P. A. Scarf, J. Electrochem. Soc., 1990, **137**, 64.
79. F. Blekkenhorst, G. M. Ferrari, C. J. Van der Wekken and F. P. IJsseling, Br. Corros. J., 1986, **21**,163 and 1988, **23**,165.
80. M. B. Rockel and M. Renner, Proc. 8th Europ. Corros. Congr., Nice, 1985, 19–21 Nov., paper 76-1.
81. P. Lau and S. Bernhardsson, Corrosion NACE, Boston, Massachusetts, 25–29 March, 1985, paper 64.
82. B. Baroux, Materials Science Forum, 1986, **8**, 91.
83. S. Lagerberg, S. Berhardsson and P. Lau, Proc. 10th Scandinavian Corr. Congress, Stockholm, 1986, 271.
84. R. Qvarfort and E. Alfonsson, Proc. 11th Scand. Corros. Congr., Stavanger, 1989, F-79.
85. R. Qvarfort, Corros. Sci., 1989, **29**, 987.
86. T. Hakkarainen, in 'Laboratory Corrosion Tests and Standards', G .S. Haynes and R. Baboian, ed., STP 866, ASTM, Philadelphia, Pa., 1985, 91.

Microbial and Biochemical Factors Affecting the Corrosion Behaviour of Stainless Steels in Seawater

V. Scotto, M. Beggiato, G. Marcenaro and *R. Dellepiane*

Istituto per la Corrosione Marina di Metalli (ICMM-CNR), Via De Marini 6/8, 16149 Genova, Italy

Abstract

The analytical characterisation of biofilms coupled with the electrochemical data from the metal substrata has been used to learn more about the mechanisms of microbially induced corrosion (MIC) on stainless steels in natural seawater. The results of various experimental tests suggest, on the whole, that bacterial presence of around 10^8 bacteria cm^{-2} and exopolymeric carbohydrate amounts (EPS) higher than 10^2 ng cm^{-2} are necessary to start MIC.

A strict correlation between EPS densities and electrochemical data measured on metal substrata was also established. The cathodic polarisation of metal or the increase of calcium disposability on surfaces increases the EPS presence in biofilm, causing a faster onset and growth of the MIC phenomenon on stainless steel. A possible MIC mechanism, useful to explain the stainless steel corrosion behaviour in natural seawater, follows, based on a critical examination of the experimental data so far collected.

1. Introduction

Studies on microbially induced corrosion are generally aimed at establishing a clear correlation between biofilm growth and the change in the metal substrata corrosion behaviour so that MIC mechanisms may be defined.

An interdisciplinary approach, with the simultaneous use of electrochemical, analytical and biological methods, is therefore necessary.

Several reviews concerning both the electrochemical methods applied in MIC studies [1, 2] and the analytical techniques used to study the nature and structure of the biofilm [3, 4] have been published recently.

The most common methods used for biofilm studies are:

- the direct measurement of the pH and dissolved oxygen inside the biofilms [5];
- the evaluation of bacterial biomass by plating suitably diluted biofilm aliquots and counting the colonies [3];
- the examination and counting of the micro-organisms by means of Epifluorescent Microscopy and/or Scanning Electron Microscopy (SEM) [6, 7];
- the characterisation of the biofilm components through non-destructive analyses like Fourier Transform Infrared Spectroscopy [8];
- the identification of the specific fatty acids utilised as biomarkers of particular slime organisms [9];
- the use of analytical methods for the evaluation of particular biofilm components [10, 11].

The ICMM team in particular has used the analytical characterisation of the biofilms, coupled with the electrochemical data coming from the metal substrata, to clarify how the corrosion behaviour of stainless steels and other active–passive alloys (such as Ti, Ni–Cr, Ni, Cu etc.) could be heavily modified in natural seawater by the adhesion of a microbiological film on their surfaces.

The higher corrosivity of natural seawater in comparison with that of an artificial one is a well established phenomenon [12–17] and it is generally attributed to a depolarisation of the oxygen reduction on stainless steel surfaces induced by still unknown bioproducts present in the slime.

The results of several experiments performed at the Institute for the Marine Corrosion of Metals (ICMM) on stainless steel exposed to natural seawater together with a MIC mechanism—developed from this work—are summarised here.

2. Experimental

The development with time of the oxygen reduction depolarisation process on stainless steel surfaces exposed to natural seawater can be followed by measuring the following:

(i) the oxygen reduction current density, i_c, at fixed cathodic potentials, in the range 0 to –500 mV SCE (potentiostatic tests);

(ii) the free corrosion potentials taken by stainless steel in the passive state (quasi-intensiostatic tests at a cathodic current value equal to the passive current, i.e. $i_c = i_{pass}$).

High quality 20Cr–18Ni–6Mo stainless steel was used to minimise both the risk of any transition from the passive to the active state and the analytical noise that would result from the presence of corrosion products in biofilms.

The sampling of biofilms for analyses from stainless steel surfaces was conducted in parallel with the recording of the electrochemical parameters. The living biomasses were evaluated by measuring the Electron Transport system (ETS) activity [12] and the bacterial population densities established through measurements of biofilm contents in Lipopolysaccharides (LPS), a biomarker of the gram negative bacteria presence [18].

The algae presence in biofilms was finally assessed by measuring the chlorophyll-a (Chl-a) amount [11] and the polysaccharidic fraction of the Extracellular Polymeric Substances (EPS) was analysed and compared with the Intracellular Carbohydrate content (INTRA) of the same biomass [19]. These analytical procedures, which have been described in detail elsewhere [11, 19, 20], were applied to the stainless steel samples held under open-circuit or potentiostatically polarised conditions in natural seawater in various experimental conditions e.g. of flow rate, sunlight, temperature, season, etc.

3. Results and Discussion

3.1 Is the oxygen reduction depolarisation induced by the presence of viable biomass on stainless steel surfaces?

The graphs in Fig. 1 show respectively:

(i) the oxygen reduction curves determined on stainless steel at increasing periods of immersion in natural seawater (Fig. 1(a));

(ii) the change with time of the free corrosion potentials observed in sea on stainless steel in the passive state (Fig. 1(b)), and

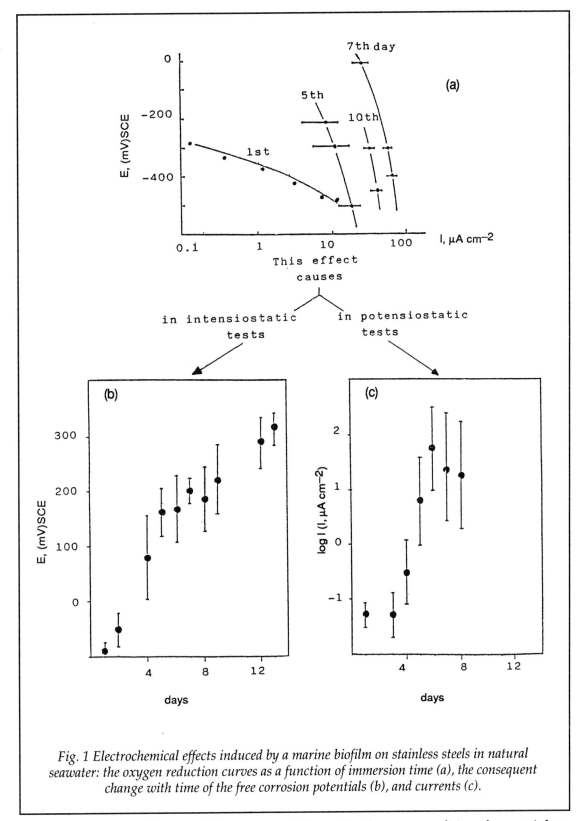

Fig. 1 Electrochemical effects induced by a marine biofilm on stainless steels in natural seawater: the oxygen reduction curves as a function of immersion time (a), the consequent change with time of the free corrosion potentials (b), and currents (c).

(iii) the increase with time of the currents delivered in the same conditions from stainless steel potentiostatically polarised at −200 mV SCE.

Both the electrochemical effects reported in Fig. 1(b) and 1(c) can be considered as logical consequences of the changes in the oxygen reduction modalities reported in Fig. 1(a).

The attribution of such depolarising effects to the settlement of marine biofilms on stainless steel surfaces follows from the similar changes with time shown [12] by the free corrosion potentials of passive stainless steel and the amounts of viable biomass, measured as ETS activity, on their surfaces (Fig. 2(a) and (b). Consequently, any external action aimed at:

- reducing the biofilm growth rate, e.g. by increasing the seawater velocity [21];
- avoiding biofilm formation, e.g. by sterilising the seawater [14];
- killing the biomass already present, e.g. by increasing the seawater temperature [22] or by adding a poison to the waters [14];

will lead to an attenuation (or disappearance) of the oxygen reduction depolarisation on stainless steel surfaces.

3.2 Is the presence of algae in the biofilm fundamental?

The influence of algae on oxygen reduction characteristics was studied by testing the corrosion behaviour of stainless steel in seawater in presence or absence of light.

The maximum respiratory activity (ETS activity) and the Chl-a contents were established in these two experimental conditions and used to characterise the biofilm nature.

The analytical data are reported in Fig. 3, where the open squares refer to tests in light and the black squares to dark exposures.

The relative positions of the ETS and Chl-a data confirm the settlement of two very different ecosystems on the surface, dominated respectively by oxygen-consuming populations or by oxygen producer organisms according to the lighting conditions in the tests.

A dominant algae population in the biofilm was in fact characterised, according to the results of Packard [23] that were established on plankton communities and since verified in ICMM on algal cultures [24], by an ETS/Chl-a ratio close to 3.28 (straight line in Fig. 3).

All previous tests indicated the same oxygen reduction depolarisation effects, independently of the algae importance in biofilm.

It follows, therefore, that bacteria, which are always present in biofilms, are highly likely to be involved in the phenomenon.

3.3 Does a minimum bacterial density exist which is necessary to induce oxygen reduction depolarisation on stainless steel surfaces?

This question was solved by comparing the LPS contents and the free corrosion potentials measured in time on stainless steel surfaces exposed to natural seawater renewed continuously or at weekly intervals.

The biomasses and the free corrosion potentials showed different growth rates (Fig. 4(a) and 4(b), on p.27), but in both cases the passive stainless steels exceeded the maximum corrosion potentials reached in sterile seawater (the E_s value in Fig. 4(a) and 4(b)) only when the LPS amounts on the surfaces achieved densities close to 10^2 ng cm^{-2} (Fig. 4(a') and 4(b')). Taking into account the fact that the mean LPS content of a marine bacteria is close to 2×10^{-6} ng/cell [25], it follows that the change of the oxygen reduction kinetics starts whenever the number of settled bacteria exceeds a threshold value of *ca.* 10^8 bacteria cm^{-2}.

On the other hand, the same data in Fig. 4 show that, once the threshold bacteria density has been reached, any subsequent increase in the free corrosion potentials cannot be associated with a proportional increase of the bacterial population.

In other words, the settlement of a minimum bacteria density on the surfaces seems to act only as a trigger for a progressive oxygen reduction depolarisation.

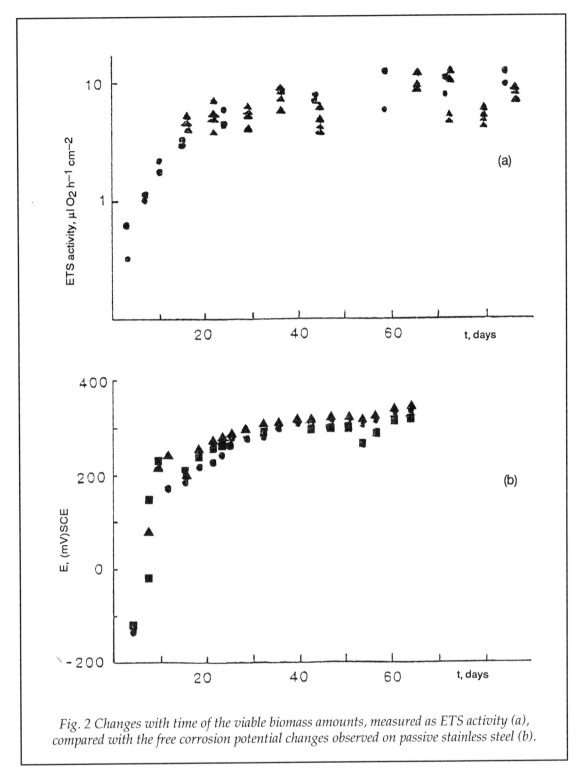

Fig. 2 Changes with time of the viable biomass amounts, measured as ETS activity (a), compared with the free corrosion potential changes observed on passive stainless steel (b).

3.4 Is the development of MIC linked to a gradual increase of bacterial EPS amounts on the stainless steel surfaces?

A series of stainless steel specimens was exposed to natural seawater both in freely corroding conditions (open circuit) and polarised at –200 mV SCE. The biofilms, collected during the

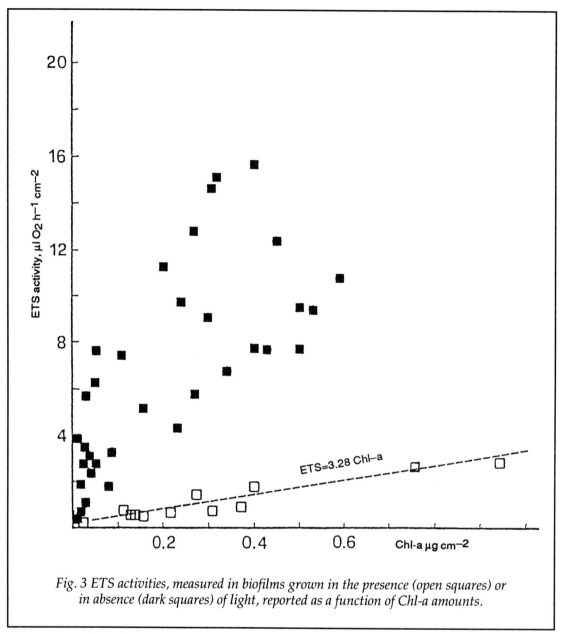

Fig. 3 ETS activities, measured in biofilms grown in the presence (open squares) or in absence (dark squares) of light, reported as a function of Chl-a amounts.

tests, were analysed to evaluate extra- and intracellular saccharidic components. The potential and current data (the letter normalised against the limiting current values I_L) are reported as functions of the EPS contents in Figs. 5 and 6 (p.28).

These Figures demonstrate the existence of a strict correlation between the EPS contents and electrochemical parameters of growth, starting from the achievement of an EPS threshold amount of about 10^2 ng cm^{-2}. These data ascribe a determining role of the MIC control to the EPS substances that always appear in strict relationship with the intracellular biomass until the completion of the electrochemical phenomenon (Figs. 7 and 8 on p.29).

Nevertheless, the EPS/INTRA ratio changed significantly according to whether the metal substrata were potentiostatically polarised at –200 mV SCE (Fig. 7) or not (Fig. 8).

These last data suggest that a cathodic polarisation of metal substrata should increase the EPS amounts in the biofilms, the intracellular biomasses being equal. The cathodic polarisation

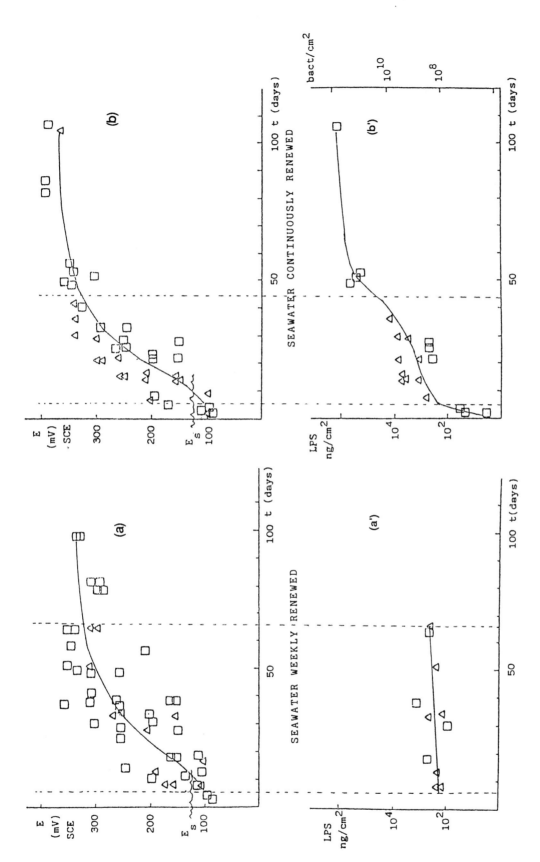

Fig. 4 *Change with time of the free corrosion potentials and LPS amounts measured on stainless steel exposed in seawater renewed weekly (4a, 4a') or continuously (4b, 4b').*

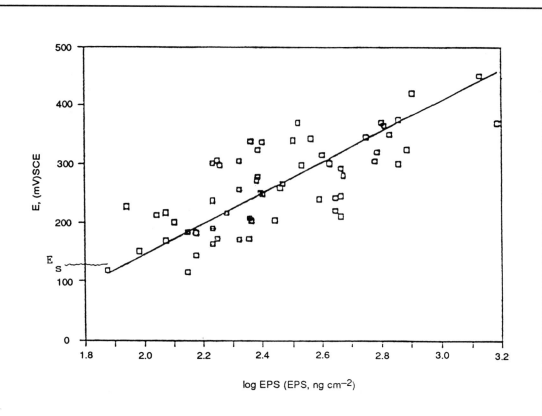

Fig. 5 Free corrosion potentials as a function of EPS amounts in biofilms.

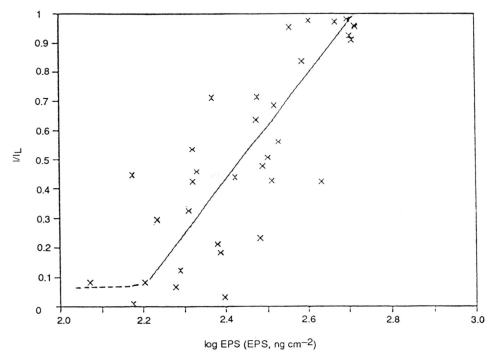

Fig. 6 Current densities, normalized against the limiting current values I_L and measured on stainless steel potentiostatically polarised at –200 mV SCE, reported as a function of EPS amount.

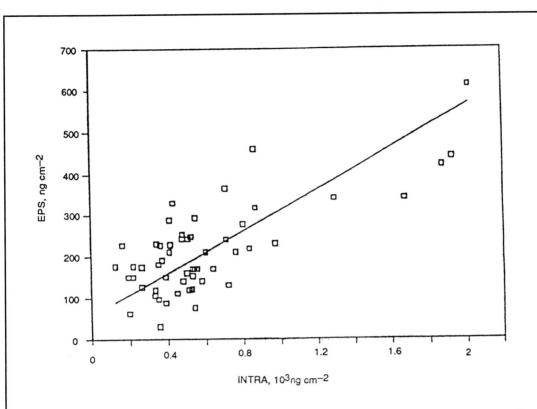

Fig. 7 EPS amounts as a function of the INTRA cellular carbohydrate contents, measured in slimes collected on stainless steel freely exposed to natural seawater.

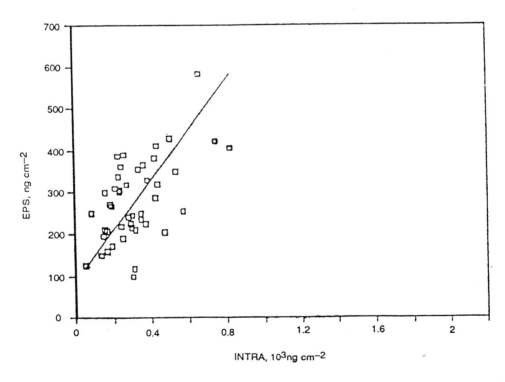

Fig. 8 EPS amounts as a function of the INTRA cellular carbohydrate contents, measured in slimes collected on stainless steel potentiostatically polarised at −200 mV SCE.

seems, in fact, to promote the adhesion of EPS to the metal surfaces—perhaps through a local enrichment of the medium by divalent ions, such as calcium or magnesium, attracted to metal surfaces by the electric fields, and promoting the gelling of the polysaccharide matrix.

To test the validity of this hypothesis, stainless steel specimens were freely exposed both in untreated seawater (Ca^{2+} = 0.013 M) and artificially enriched in calcium up to 0.018 M. In the course of the tests, the free corrosion potentials of the stainless steel were measured and biofilm samples collected and analysed as before.

The EPS and INTRA amounts, measured in the presence of calcium excess, are reported in Fig. 9 and appear in perfect agreement with the data previously obtained on samples cathodically polarised in untreated seawater (line a in Fig. 9). Furthermore, the open circuit corrosion potentials increased more quickly in excess of calcium (black squares in Fig. 10) than in untreated seawater (black triangles in the same Figure) signalling that the faster the EPS accumulation is on stainless steel surfaces, the faster occurs the depolarisation of the oxygen reduction.

3.5 Is the EPS presence alone sufficient to cause MIC?

It has been well established [14] that the depolarising effects on oxygen reduction can be eliminated by adding sodium azide, NaN_3, to the seawater. Considering that this product is an enzymic inhibitor of the respiratory chain and unable to detach biofilms from surfaces, the sharp decrease in potential that has been observed on its addition to the medium, indicate that, in this case, the EPS presence alone on surfaces is not sufficient to cause MIC.

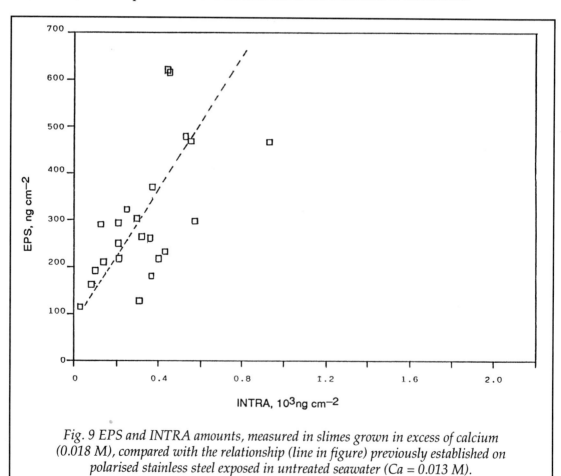

Fig. 9 EPS and INTRA amounts, measured in slimes grown in excess of calcium (0.018 M), compared with the relationship (line in figure) previously established on polarised stainless steel exposed in untreated seawater (Ca = 0.013 M).

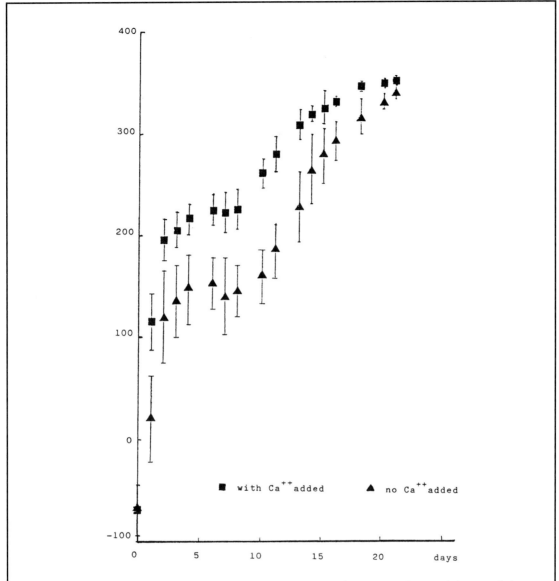

Fig. 10 Change with time of the free corrosion potentials, measured on stainless steel, in presence (black squares) or in absence (black triangles) of calcium added to seawater.

In our opinion, all the data that have been presented suggest an MIC mechanism for stainless steel in seawater as is schematically represented in Fig. 11.

According to this scheme, the settlement of about 10^8 bacteria cm^{-2} on the surfaces is necessary for the production of extracellular polymers (EPS) in a sufficient amount to immobilise, firmly and evenly, viable cells and exudates on metal and to accumulate on the metal surfaces effective concentrations of still unknown bioproducts that will catalyse the oxygen reduction.

The immobilising efficiency of EPS is improved in the presence of higher calcium concentrations in the medium or near the metal surfaces which promote EPS gelling. In our opinion, EPS has a necessary but not sufficient role in causing MIC if it is not coupled with the presence in the biofilm of some biocatalysts, perhaps belonging to the Electron Transport System of the respiratory chain, which affect oxygen reduction kinetics.

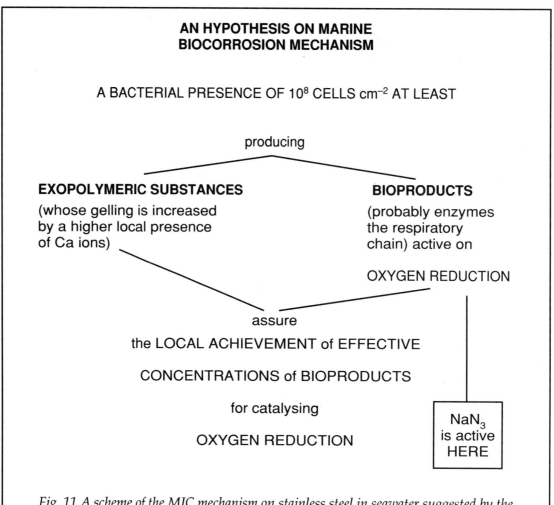

Fig. 11 A scheme of the MIC mechanism on stainless steel in seawater suggested by the experimental results.

4. Conclusion

Electrochemical and analytical methods were used together to establish whether a connection exists between biofilm growth and the oxygen reduction depolarisation observed on stainless steel surfaces in seawater.

The results of several experimental tests suggest that:

- the oxygen reduction kinetics are not influenced by the presence of algae in the biofilm;
- the onset of the oxygen reduction depolarisation effects requires the presence on stainless steel surfaces of 10^8 bacteria cm^{-2} and an EPS amount higher than about 100–150 ng cm^{-2};
- the faster is the accumulation of EPS on a surface, the faster the oxygen reduction depolarisation occurs;
- the accumulation of EPS on surfaces is facilitated by cathodic polarisation of metal substrata;
- the presence, in the medium or near to the metal surfaces, of higher contents of calcium ions appears to increase the EPS fraction in the biofilm, accelerating the occurence of corrosive situations;

- bacterial products, so far unidentified, acting as biocatalysts of oxygen reduction will probably reach effective concentrations on metal surfaces, as a result of EPS immobilising effects.

References

1. S. C. Dexter, D. J. Duquette, O. W. Siebert and H. A. Videla, Corrosion, 1991 **47**, (4), 308–318.
2. F. Mansfeld and B. Little, Corros. Sci., 1991, **32**, (3), 247-272.
3. D. Y .C. Fung, Biodeterioration 7, pp.647-656, D. R. H. Hoghton, R. N. Smith and H. O. W. Eggins, Eds, Elsevier Applied Sciences, 1988.
4. C. Gaylarde and P. Cook, Biodeterioration 7, pp.657-663; D.R.H.Hoghton, R.N.Smith and H.O.W.Eggins, Ed., Elsevier Applied Sciences, 1988.
5. Z. Lewandowski, W. C. Lee, W. G. Characklis and B. Little, Corrosion, 1989, **45**, (2), 92–98.
6. J. S. Maki, B. J. Little, P. Wagner and R. Mitchell, Biofouling, 1990, **2**, 27–38.
7. D. S. Marszalek, M. Sol, L. Gerchakov and R. Udey, Applied and Environmental Microbiology, 1979, **38**,(5), 987–995.
8. D. E. Nivens, P. D. Nichols, J. M. Henson, G. G. Geesey and D. C. White, Corrosion, 1986, **42**, (4), 204–210.
9. R. H. Findlay and D. C. White, Applied Environmental Microbiology, 1983, **45**, 71.
10. H. M. Lappin-Scott and W. Costerton, Biofouling, 1989, **1**, 323–342.
11. V. Scotto, G. Alabiso and G. Marcenaro, Bioelectrochemistry and Bioenergetics, 1986, **16**, 347–355.
12. A. Mollica et al., Proc. 6th Int. Congr. on Marine Corrosion and Fouling, Athens, 1984, pp.269–281.
13. R. Johnsen and C. Bardal, Corrosion, 1985, 41, (5), 296.
14. V. Scotto, R. Di Cintio and G. Marcenaro, Corros. Sci., 1985, **25**, 185.
15. S. C. Dexter and G. Y.Gao, CorrosionNACE '87, San Francisco, California, 9–13 March, 1987, paper 377.
16. G. J. Brankevich, M. L. F.De Mele and H. A. Videla, Marine Techn. Soc. J., 1990, **24**, (3), 18–28.
17. T. Ford and R. Mitchell, The Ecology of Microbial Corrosion, pp.231–262, K. C. Marshall, Ed. Plenum Press, 1990.
18. V. Scotto J. Guezennec and G. Alabiso, Poster presented at the International Conference of Electrified Interfaces (Solid/electrolyte and Biological Systems), Bologna, Sept., 1988.
19. V. Scotto, G. Alabiso, M. Beggiato, G. Marcenaro and J. Guezennec, Proc. 5th Europ.Congr. on Biotechnology, Copenhagen, Vol.2, pp.866–871, July 1990, C. Christiansen, L. Munck and J. Villadsen, (eds). Munksgaard Int. Publ., Copenhagen.
20. G. Alabiso, V. Scotto and G. Marcenaro, Nova Thalassia, Vol. 6, suppl.1983-1984, pp.451–457.
21. A. Mollica and A. Trevis, Proc. 4th Int. Congr. on Marine Corrosion and Fouling, Juan les Pins, pp.251-265, 1976, edited by Centre de Recherches et Etudes Oceanographiques, Boulogne, France.
22. A. Mollica, A. Trevis, E. Ventura, G. Traverso, G. Decarolis and R. Dellepiane, Corrosion, 1988, **44**, (4), pp.194–198.
23. T. T. Packard, D. Harmon and J. Boucher, Tethys, 1974, **6**, (1-2), 213.
24. G. Alabiso, G. Marcenaro and V. Scotto, Oebalia 1985, 11, pp. 853–856.
25. M. Maeda et al., Marine Biology, 1983, **76**, 257–262.

Experience

Marine Corrosion Tests on Metal Alloys in Antarctica: Preliminary Results

G. ALABISO*, U. MONTINI, A. MOLLICA, M. BEGGIATO, V. SCOTTO,
G. MARCENARO AND R. DELLEPIANE

Istituto per la Corrosione Marina dei Metalli (ICMM-CNR), Via De Marini 6/8, 16149 Genova, Italy
*Istituto Talassografico Taranto, Via Roma 3, 74100 Taranto, Italy

Abstract

The aim of this research was to study the corrosivity of the Antarctic marine environment on a series of metal alloys of wide industrial use.

In this preliminary note some results concerning the corrosion behaviour of stainless steels and aluminium alloys are reported. Particular attention was given to establish a correlation between biofilm growth and corrosion on stainless steels, and to study the influence of temperature on aluminium alloy corrosion.

The research was supported by the Italian National Antarctic Research Programme and was carried out in the Italian Base situated in Terra Nova Bay (Ross Sea, Victoria Land) during the 1989/90 and 1990/91 campaigns.

During the two campaigns short lab-tests, (*ca.* 2 months duration) were carried out.

During the 1989/90 campaign a loading structure with 120 metal samples was immersed into the sea at 50m depth and recovered during the 1990/91 campaign.

In the case of stainless steels, the comparison with the data obtained in other seas shows that:

- the presence of biofilms on metal surfaces in Antarctic seawater is less dangerous with respect to the probability of initiation of crevice corrosion than in other seas;
- if localised corrosion begins, the corrosion rate is close to that observed in other seas;
- in the Antarctica seas, like in other seas, an artificial rise of sea water temperature by 25°C over the mean local values is sufficient to strongly delay the microbiological influences on the corrosion processes.

In the case of aluminium alloys, the data show that (i) at low temperatures, under 10°C, the oxygen reduction sustains high localised corrosion rates; and (ii) over 10°C, low uniform corrosion rates are sustained by the hydrogen reduction.

1. Introduction

Areas with extreme climatic conditions, such as the polar regions, are being increasingly used and therefore it becomes necessary to provide a more direct and detailed knowledge of metal alloys behaviour with regard to marine corrosion [1–5].

Moreover, only a few researchers, exclusively in the North Sea, have studied the influence of marine biofilm on corrosion in these areas [6].

The aim of this research was to study the Antarctic marine environment corrosivity on a series of metal alloys with wide industrial uses.

Results of the corrosion behaviour of stainless steels and aluminium alloys are reported, with particular attention given to establishing a correlation between biofilm growth and corrosion on stainless steels, and pointing out the influence of temperature on aluminium alloys corrosion behaviour. The research was supported by the Italian National Antarctic Research Programme and was carried out in the Italian Base situated in Terra Nova Bay (Ross Sea, Victoria Land) Lat.: 74°41' 42"S, Long.: 164°07' 23"E (Fig. 1), during the 1989/90 and 1990/91 Campaigns.

2. Materials and Methods

The alloys examined were as follows: stainless steels type AISI 304, 316 and 20Cr–18Ni–6Mo, copper, brass, cupro–nickel 70/30, Monel, aluminium and aluminium–3.5 magnesium alloy.

During the two campaigns short laboratory tests (*ca.* 2 months long), and a field test, one year long, were carried out in Terra Nova Bay.

2.1 Laboratory tests

At the Base, in a suitably equipped room, 200 specimens (including stainless steels AISI 304, 316, 20Cr–18Ni–6Mo, Aluminium and Al–3.5Mg alloy) were exposed in four tanks, each containing 70 l, of seawater continuously renewed at the rate of 2 l min^{-1} (Fig. 2).

The first tank was maintained at a temperature close to that of natural seawater (*ca.* 2°C), whereas the others were warmed up to 10, 20 and 30°C respectively.

Open-circuit corrosion potentials were followed with time and polarisation curves also obtained.

Changes in oxygen reduction behaviour on stainless steels were recorded with respect to time and temperature, using specimens of stainless steel galvanically coupled to mild steel samples through a resistor.

Galvanic currents were determined from the potential drop on the resistors and used for calculating the oxygen reduction currents (i_c) on the surface of stainless steels; at the same time, the corrosion potentials (E_c) taken up by the stainless steels were measured.

By changing the resistors (10, 100, 330 and 1000 Ω respectively), the different points P (corresponding to i_c and E_c values) on the oxygen reduction curves were obtained (Fig. 3, p.39).

Biofilm samples, grown at different temperature conditions on the stainless steel surfaces, were periodically collected and freeze dried for subsequent analytical characterisation in Italy. This analysis involved the measurement of proteins, carbohydrates, lipopolysaccharides and chlorophyll-a contents and the reconstruction of fatty acid profiles.

Biofilm samples were also spread, in Antarctica, onto marine Agar and incubated at 4°C.

In Italy the bacterial strains were isolated and purified for taxonomical purposes.

During the 1989/90 Campaign, 72 bacterial strains were isolated and purified.

2.2 Field test

In addition to the laboratory tests, in the 1989/90 Campaign a loading structure with 120 metal samples was immersed into the sea at 50m depth and recovered during the 1990/91 Campaign. Sea-water temperature was between –1.8 and 0°C during the exposure and all the above alloys were used in this test.

The structure was equipped with an acquisition data system to measure free corrosion potentials and corrosion currents. In the case of AISI 304 and 316 samples, these measurements indicated both the time to crevice corrosion initiation, and the corrosion currents, as shown schematically in Fig. 4 (p.39).

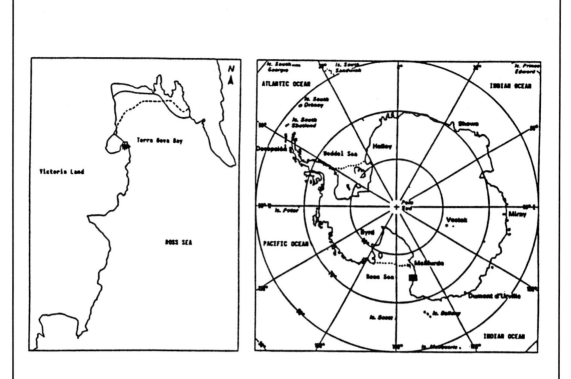

Fig. 1 The Italian permanent summer station at Terra Nova Bay, Ross Sea, Antarctica.

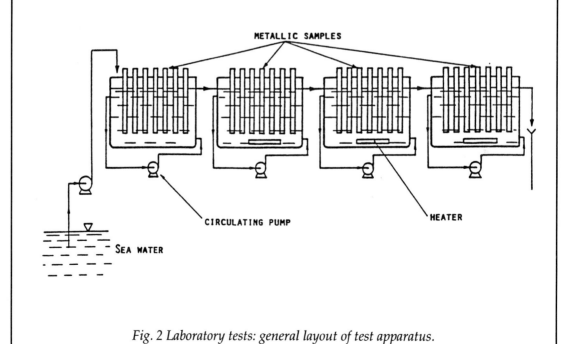

Fig. 2 Laboratory tests: general layout of test apparatus.

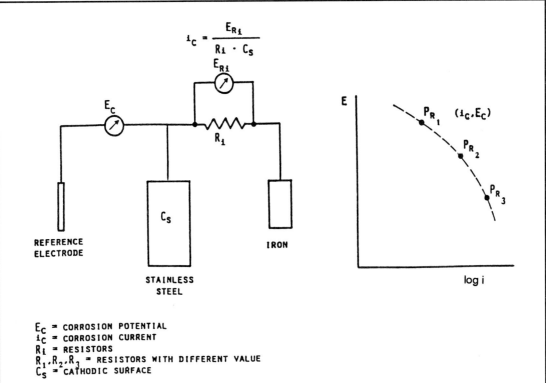

Fig. 3 Assembly of the stainless steel samples galvanically coupled to mild steel samples.

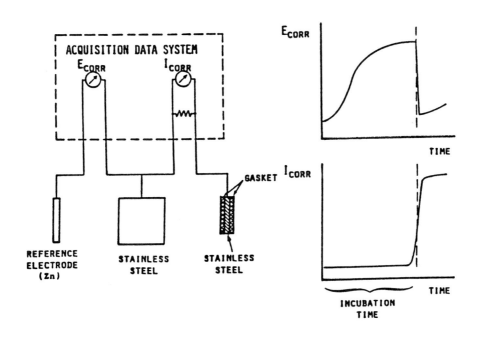

Fig. 4 Field tests: general layout of test apparatus.

3. Results and Discussion

3.1 Stainless steels

The aim of the laboratory and field tests was to establish if oxygen reduction depolarisation, induced by biofilm growth on metal surface, could be observed in Antarctic water as well as in the Mediterranean and in other seas.

This phenomenon affects both the corrosion initiation probability and corrosion rate [7–13].

Figure 5 shows the shift of the oxygen reduction curve vs time at 2°C in Antarctica.

A strong change of the cathodic curve, followed by stabilisation, can be observed between the 10th and the 20th days.

Although the results of biological and biochemical analysis are not yet available, the *in situ* observation showed that biofilm formation became clearly evident on the metal surface from the 3rd week of exposure.

Nevertheless, some differences exist between the shape of cathodic polarisation curves obtained in Antarctica after stabilisation and those observed, in similar experimental conditions, in other seas.

In Fig. 6 the final cathodic curve obtained in the Ross Sea and a cathodic curve obtained in similar experimental conditions in the Mediterranean Sea are plotted.

Two schematic anodic polarisation curves describing, repetitively, a stainless steel in active and passive state are added.

Two remarks, concerning free corrosion potentials in the passive state and corrosion rates in the active state respectively, can be made:

(i) In the Ross Sea a free corrosion potential in the passive state close to zero mV SCE can be expected: this value is much lower than the free corrosion potential observed in the Mediterranean Sea (300–400 mV SCE).

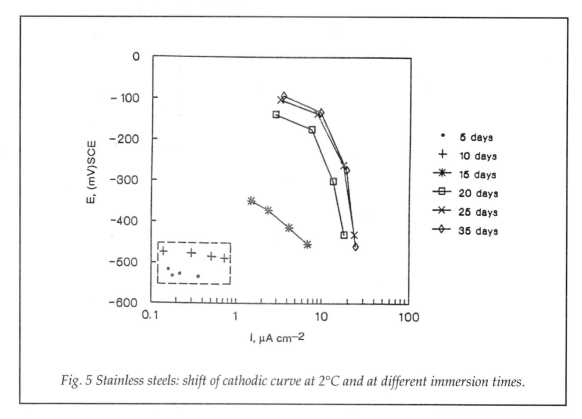

Fig. 5 Stainless steels: shift of cathodic curve at 2°C and at different immersion times.

The graphs in Fig. 7 show the trend of free corrosion potentials measured on the stainless steels exposed at 2°C in the laboratory.

In accordance with expected results and for every stainless steel examined, the free corrosion potentials reached maximum values between 0 and 50 mV SCE after about two exposure weeks.

Looking at the stability of free corrosion potential we can observe that:

- this low level was not sufficient to initiate any form of localised corrosion on the higher quality stainless steel (20Cr–18Ni–6Mo);
- only a few AISI 316 samples seemed to show a transition from passive to active state at the end of the test;
- a significant number of negative going potential excursions signals a high probability of localised corrosion on AISI 304 samples.

(ii) The graph of Fig. 6 shows that in Antarctica the localised corrosion propagation rate is similar to that observed in the Mediterranean Sea: localised corrosion rate is sustained by cathodic current of at least 1 μA cm^{-2}.

In conclusion, the lab-tests suggest that, in comparison with the other seas, in the Ross Sea:

- the probability of localised corrosion initiation is lower;
- the corrosion rate, if localised corrosion starts, is similar.

Biofilm growth effect on corrosion initiation probability and on corrosion rate should also have been studied by means of the long-term field test to confirm laboratory conclusions.

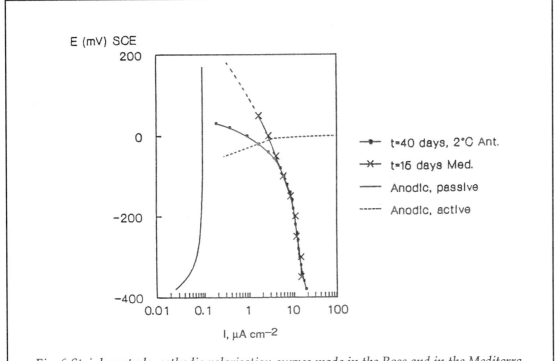

Fig. 6 Stainless steels: cathodic polarisation curves made in the Ross and in the Mediterranean Seas; schematic anodic behaviour of a stainless steel in passive and active state.

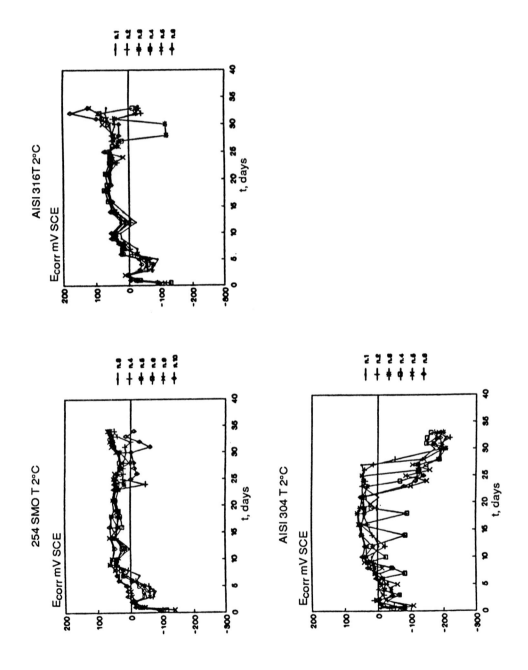

Fig. 7 Stainless steels: free corrosion potentials vs time.

Unfortunately, during the exposure the acquisition data system was seriously damaged owing to water infiltration and no measurements are available.

Consequently, it was only possible to determine the weight loss sustained by each sample during the immersion year. Nevertheless, an attempt was made to evaluate a possible distribution of crevice corrosion incubation time using the following system.

For instance, in the case of the AISI 304 (Table 1):

- every sample weight loss was scheduled in decreasing order;
- incubation time of the sample which showed the higher weight loss was arbitarily reported as zero, and therefore a propagation time equal to 12 months could be attributed;
- mean weight loss per month ($\overline{\Delta Pm}$) was obtained when crevice corrosion started;
- the weight loss of the remaining samples was considered proportional to the corrosion propagation time;
- corrosion propagation time for the other samples was calculated as

$$\text{Propagation time} = \Delta P / \overline{\Delta Pm}$$

- incubation times were calculated as

$$T_{incub} = 12 - T_{prop}$$

So, the survival vs time curves, for AISI 304 and AISI 316, are represented in Fig. 8.

A survival-time curve, obtained in the Mediterranean Sea for a similar crevice configuration on AISI 316 [14], is reported in the same figure.

In spite of the qualitative value of the former, the comparison between the two graphs shows that the initiation probability is at least one order of magnitude less in the Ross Sea than in the Mediterranean Sea. (Note that units reported on x-axis are months for the Ross Sea and days for the Mediterranean Sea.)

According to the results of laboratory tests, although the probability of initiation of localised

Table 1 Calculations of AISI 304 crevice corrosion incubation time

ΔP (g)	Propagation time (months)	Incubation time (months)	Survival (%)
2.17	12.0	0.0	87.5
1.92	8.6	3.4	75.0
1.82	8.1	3.9	62.5
1.71	7.6	4.4	50.0
1.46	6.5	5.5	37.0
1.40	6.2	5.8	25.0
0.17	0.8	11.2	12.5
0.16	0.7	11.3	0.0

$\overline{\Delta Pm}$ = 0.18 (g)

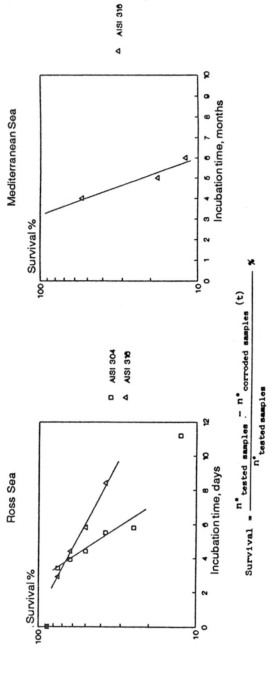

Fig. 8 Stainless steels: comparison between survival before crevice corrosion in the Ross and in the Mediterranean Seas. (note different time scales).

$$\text{Survival} = \frac{n° \text{ tested samples} - n° \text{ corroded samples}}{n° \text{ tested samples}}(t) \cdot \%$$

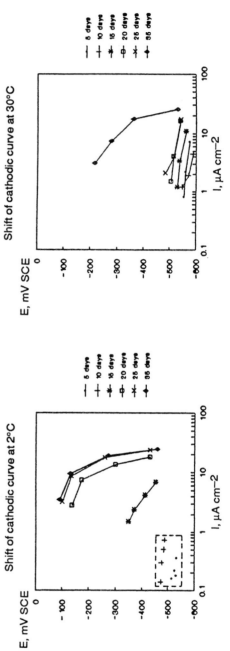

Fig. 9 Stainless steels: comparison between the cathodic curves shift obtained at 2 and 30°C.

corrosion is lower, the corrosion rate is high once the corrosion starts: for example, 2mm-thick samples were found to be perforated at the end of the test.

The final observation concerns the effect of rise in temperature. In Fig. 9 the shift of the cathodic curve observed at 30°C is shown in comparison with the shift observed at 2°C.

The shift at 30°C is very delayed: this is in agreement with the observations made in other seas, in respect of the reduction of microbial interference on cathodic behaviour, when temperature is increased by 25–30°C above the natural seawater temperature [15,16].

3.2 Aluminium alloys

Figure 10 shows free corrosion potentials of aluminium samples recorded against time during laboratory tests at 2, 10, 20 and 30°C.

The trend of the curves shows that below 10°C the corrosion potentials ennoble rapidly, reaching stable values between –700 and –750 mV SCE. At the end of the exposure period, all the samples exhibited signs of localised corrosion.

Above 10°C the corrosion potentials, after an initial ennoblement, fell to less noble values (i.e. below –850 mV SCE). At the end of the experiment, no signs of corrosion were present.

Polarisation curves, obtained at the end of the tests (Fig. 11), agree with the previous results and suggest the following conclusions:

-At low temperature, an oxygen reduction rate of some $\mu A\ cm^{-2}$ sustains corrosion in a potential range which makes localised corrosion initiation possible.
-At the higher temperature, a hydrogen reduction rate of some tenths of $\mu A\ cm^{-2}$ sustains corrosion in a potential range in which the alloy is in the passive state.

The behaviour of Al–3.5Mg alloy was very similar to that of aluminium.

The samples recovered after one year of immersion in the Ross Sea at an annual mean temperature of about –1°C confirmed the results obtained in the laboratory tests at 2°C.

The prediction was that they should have supported a corrosion rate in the form of localised corrosion of about 2–4 $\mu A\ cm^{-2}$, equivalent to a total weight loss of 2.5–5g/year for a 450 cm immersion.

In agreement with this prediction, the average effective weight loss was 3.9 ± 0.62g for Al–Mg alloy samples and 4.68 ± 0.77 g for aluminium samples. All samples showed localised corrosion, in the form of crevices and pits which were predominantly on the sample edge. Frequently, the crevice corrosion was sufficiently intense to perforate the 2 mm-thick samples.

4. Conclusions

Field and laboratory tests conducted on stainless steels (AISI 304, AISI 316, 20Cr–18Ni–6Mo) and aluminium alloys (aluminium and Al–3.5Mg) in Antarctica allow the following, preliminary conclusions to be drawn:

• Oxygen reduction depolarisation induced by biofilm growth on the surface of stainless steels, similarly to that observed in other seas, was also found in Antarctica with a seawater temperature close to 0°C.

• Nevertheless, in comparison with the Mediterranean Sea, some differences in the final shape of cathodic curves, when the surface of stainless steels has been covered by biofilms, can be observed.

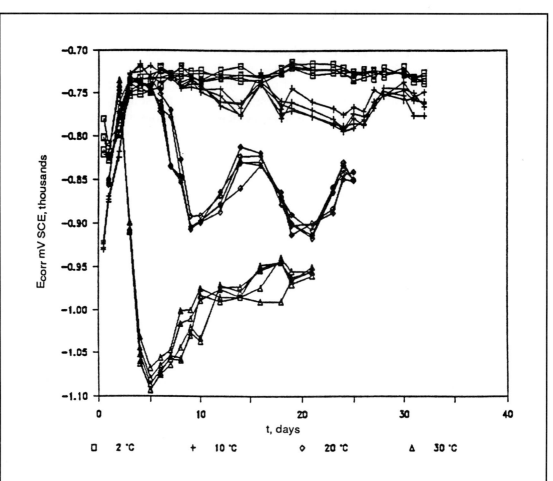

Fig. 10 Aluminium: free corrosion potentials vs time measured at 2, 10, 20 and 30°C.

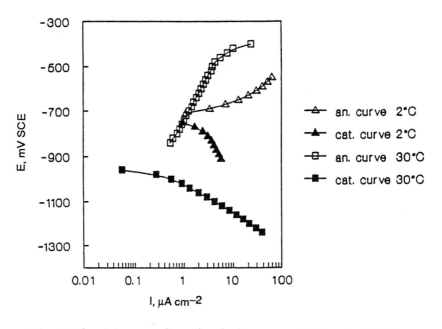

Fig. 11 Aluminium: anodic and cathodic curves at 2°C and at 30°C.

The differences point to a decrease in the probability of localised corrosion initiation in Antarctica; although once started the rate of propagation of localised corrosion in the two regions is about the same.

- A rise in temperature above 30°C tends to delay sharply the oxygen reduction depolarisation induced by biofilm growth on the surface of stainless steels.

- The corrosion of aluminium alloys is heavily affected by the seawater temperature: thus, under 10°C localised corrosion, in the form both of crevice and pitting corrosion, is easily enhanced with the propagation sustained by the oxygen reduction. Above 10°C only uniform corrosion, sustained by hydrogen reduction, occurs.

5. Acknowledgement

The authors thank very much ENEA logistics for their great help to the group activity during the Antarctic Campaigns.

References

1. I. Smuga-Otto, Corrosion NACE '86, Houston, Tx, March 1986, p.57.
2. D. G. Lignau, *ibid.*, p.57.
3. A. C. Madsen, F.O. Mueller, R. Geisert, J. B. Lee and J. Choi, Materials Performance, 1986, **26**,(2), p.49.
4. C. M. Schillmoller and B. D. Craig, Materials Performance, 1987, **27** (10), 46.
5. G. Alabiso, U. Montini, V. Scotto, A. Mollica, G. Marcenaro, M. Beggiato, R. Dellepine, R. Bandelloni and D. Tiberti, Nat. Sci. Com. Ant. Ocean. Camp. 1989–90, Data Report, 1991, pp. 395–409.
6. R. J. Edyvean, L. A. Terry and G. B. Picken, Int. Biodeterior. Bull., 1985, **21**, 4, 277.
7. A. Mollica, A. Trevis, E. Traverso, G. Ventura, V. Scotto, G. Alabiso, G. Marcenaro, U. Montini, G. de Carolis and R. Dellepiane, Proc. 6th Int. Congr. on Marine Corrosion and Fouling, Athens, 5–8 Sept. 1984, p.269.
8. V. Scotto, G. Alabiso and G. Marcenaro, Bioelectrochemistry and Bioenergetics, 1986, **16**, 347.
9. R. Johnsen and E. Bardal, Corrosion NACE '85, Vol. 41, p.286.
10. A. Desestret, Materiaux et Techniques, 1986, **7/8**, 31.
11. J. M. Krougman and F. P. IJsseling, Proc. 5th Int. Cong. on Marine Corrosion and Fouling: Marine Corrosion, Barcelona, Spain, May 1980, p.214.
12. R. Holte, E. Bardal and P.O. Gartland, Corrosion '88, NACE, St Louis, MO, paper 93.
13. S. C. Dexter and G. Y. Gao, CorrosionNACE '87, Houston, TX, paper 77.
14. A. Mollica, Materiaux et Techniques "Special" Biocorrosion, Dec. 1990, p.17.
15. A. Mollica, A. Trevis, E. Traverso, G. Ventura, G. de Carolis and R. Dellepiane, Corrosion '89, NACE, January 1989, Vol. 45p. 48.
16. T. S. Lee, R. M. Kain and J. W. Oldfield, Materials Performance, 1984, **24**, 9.

4

North Sea Experience with the Use of Stainless Steels in Seawater Applications

R. JOHNSEN

Statoil Research Centre, Postuttak, N-7004 Trondheim, Norway

Abstract

Seawater is one of the most corrosive environments for many metals and alloys, including active/passive alloys such as stainless steel. During the last 10–20 years a number of different stainless steel alloys have been developed mainly for seawater applications.

This paper describes the experience from the use of different austenitic and duplex (ferritic–austenitic) stainless steels in seawater systems in connection with oil and gas production in the Norwegian sector of the North Sea. Some case histories are given and results from R&D projects on the same topic are briefly described.

The design philosophy used by Norwegian oil companies when selecting alloys for seawater systems is given, based on practical experience and results from R&D projects, and other materials used for seawater piping systems besides stainless steels are also considered.

1. Introduction

When oil extraction started in the Norwegian sector of the North Sea in the early sixties, the operating responsibility was given to American oil companies. Phillips Petroleum Company was selected as operator for the Ekofisk Field, whereas Mobil Exploration was operator for the Statfjord Field. These companies, having gained experience in oil production around the world mainly from onshore fields, brought their experience to the North Sea development [1, 2].

Cement-lined carbon steel was the main choice of construction material for seawater piping systems, while copper-based alloys were selected for other parts of the system. For valves, rubber-lined carbon steel, Ni-resist and copper-based alloys were most widely used.

Leaks in the piping systems were experienced soon after system start-up—particularly in cement-lined carbon steel pipes. The leaks were often repaired temporarily with patches, or larger pipe sections were replaced. A large inspection programme at the Statfjord A platform in 1982 showed a lot of corrosion attack in cement-lined pipes. The attack was mainly concentrated in field joints, in sections with turbulence or high velocity, in sections with cracks or defects in cement-lining and where metallic couplings could initiate galvanic corrosion.

2. The Use of Stainless Steels in Seawater Systems

As failures were being reported from the seawater handling systems in the North Sea, stainless steel producers were bringing new alloys into the market. Among these, austenitic stainless steels, Avesta 254 SMO (UNS S31254) with approximately 6% Mo had been specially developed for seawater application (see Table 1). The first installation of a small quantity of this alloy was carried out at the Frigg Field in 1979.

The selection of 254 SMO for seawater systems, including:

Table 1 Chemical analysis of some stainless steel alloys

Type	UNS number	Producer	Cr	Ni	Mo	Cu	N	PRE_N
AL-6XN	UNS NO8367	Allegheny	20.8	25	6.5	0.15	0.20	45.4
Cronifer 1925 HMO	UNS NO8925	VDM	20	25	6.4		0.20	44.3
254 SMO	UNS S31254	Avesta	20	18	6.1	0.7	0.20	43.9
Inco 25-6Mo		INCO	19.5	25.5	6.2	1.0	0.19	42.9
SAF 2507		Sandvik	25	7	4	0.2	0.30	43
Zeron 100	UNS S32760	Weir	25	7	3.5	0.8	0.25	40.6
SAF 2205	UNS S31803	Sandvik	22	5.5	3.0		0.14	34.1
AISI 316L			18	14	2.5	0.3		26.2

- ballast water
- fire water
- seawater injection (before deoxygenation)
- fresh water generation
- cooling system

on the Gullfaks A platform in 1980 was a breakthrough for the use of 6Mo-steel in the offshore industry. (The Gullfaks Field was the first oil field in the North Sea operated by a Norwegian oil company!) The order included 40 000 m of pipes from 1/2 to 36" in diameter, 5000 flanges and 23 000 fittings. The total weight of the order was approximately 700 000 kg. In addition, more than 600 valves are made from 254 SMO.

Later on, 6Mo-steel has been selected for the seawater systems at the following fields in the Norwegian sector:

- Oseberg   (Norsk Hydro)
- Snorre   (Saga Petroleum)
- Sleipner   (Statoil)
- Draugen   (Statoil - Shell)

In all these fields 6Mo-steel is the main material for all piping, fittings and valves. The seawater pumps, however, have up to now *not* been produced in 6Mo-steel. This is mainly due to the cost compared to alternative materials like nickel-aluminium-bronze or duplex stainless steel

(UNS S31803 or equal). Using the nickel-aluminium-bronze, one has to insulate carefully the pump from the 6Mo-piping to prevent galvanic corrosion of the pump or cathodically protect the pump with sacrificial anodes.

Since there has been some corrosion failures on seawater pumps made from nickel aluminium bronze, the most recently developed fields in Norway have selected more corrosive resistant material for pumps. Due to its high strength, duplex stainless steel with approximately 25% Cr, often called 'super duplex', is frequently used (Table 2). Among these are Zeron 100 from Weir and SAF 2507 from Sandvik.

On some older installations 6Mo-steels are often used as replacement material for part of the seawater system.

In addition to 6Mo-steels, duplex stainless steel (UNS S31803) and even AISI 316L are used for some equipment exposed to seawater.

During recent years stainless steel has also been selected for other applications in the North Sea to reduce corrosion failures. Some examples are shown in Table 3.

3. Experience from the Use of Stainless Steels in Seawater

The seawater in most of the systems that have been discussed in this chapter was fully oxygenated and chlorinated with a residual chlorine level in the range of 0.5–1.5 ppm.

3.1 AISI 316L

In offshore industry AISI 316L is frequently used for hydraulic lines. Both for lines on platforms (dry conditions) and for subsea lines connecting a platform and a subsea unit (wet conditions). Several failures due to localised corrosion have been reported on subsea lines. In order to prevent corrosion failures, these lines have to be cathodically protected.

Case History 1. In the water injection system at the Statfjord B platform the seawater (backwash) filters are made from AISI 316L. The seawater coming from the oil coolers enters the filter at

Table 2 Selected pump materials for some fields in the Norwegian sector of the North Sea

FIELD	OPERATED BY	PUMP MATERIAL
Gullfaks	Statoil	Ni–Al bronze
Oseberg	Norsk Hydro	Duplex stainless steel (UNS S31803) with anodes
Sleipner	Statoil	"Superduplex"*
Snorre	Saga	"Superduplex"*/ 6Mo-steel
Draugen	Shell	"Superduplex"*

*Duplex stainless steel with minimum 25% Cr.

Table 3 Examples of the use of stainless steel in oil production systems

FIELD	OPERATED BY	SYSTEM	PUMP MATERIAL
Gullfaks	Statoil	Cladding in pressure vessels	AISI 316L
		Produced water	6Mo-steel
Tommeliten	Statoil	9" (23cm) flowline	Duplex stainless steel (UNS S31803)
Snorre	Saga Petroleum	Process piping	6Mo-steel
Oseberg	Norsk Hydro	Process piping	Duplex stainless steel (UNS S31803)
Veslefrikk	Statoil	Process piping	Duplex stainless steel (UNS S32760)

a temperature in the range of 22–25°C. The seawater is still fully oxygenated and has a residual chlorine level of 0.9 ppm.

After some time in operation severe corrosion attacks were observed in flanged connections and in other parts with local crevices. To prevent this type of corrosion a cathodic protection system based on replaceable, sacrificial anodes was designed and installed based on a current density requirement of 150 mA m^{-2}.

3.2 Duplex Stainless Steel (22% Cr)

Duplex stainless steel (UNS S31803 or equivalent) suffers from localised corrosion in seawater under certain conditions even at temperatures in the range of 10–20°C. As shown in Case History 2, components made from this alloy should be cathodically protected to prevent corrosion damage.

Case History 2. On some Statoil operated platforms seawater pumps are made from duplex stainless steel according to UNS S31803. These pumps are submerged in the sea and connected to the platform deck by a hydraulic inner pipe and several pipe stacks (cargo pipes). According to the specification the inner and the outer part of the pump should be protected by sacrificial anodes.

After four years in operation some of the pumps on one platform were inspected. The internal parts of all the pumps were in good condition with no corrosion attack. According to the design nearly all the anodes were consumed.

Serious corrosion attack was found on some of the external flanges on the pipe stacks. The flanges were forged, and the maximum depth of attack was 22 mm. The inspection showed that the external parts had not been cathodically protected, as specified in the design.

3.3 6Mo-steel

After five years in service at the Gullfaks Field the experience from the use of 254 SMO in seawater system is extremely good. Some failures have been reported in welds and fittings.

Examinations have shown that a change of alloys very often has been the reason for the failures. Thus, on several occasions AISI 316L had been used instead of 254 SMO; in these cases the quality control system had not worked properly.

Corrosion failures in welds have been caused by the use of non-recommended filler metal. Such changes of alloys and filler metal has been the most serious problem when using 254 SMO.

Case History 3. In some cases, 254 SMO has suffered from corrosion in seawater. Plate type seawater evaporators for freshwater production made with titanium plates and pipework and tank wall from 254 SMO are installed on one of the Statfjord platforms. After one year in service some corrosion problems were identified on one of these units. Crevice corrosion was observed on flange seal faces made from 254 SMO in the fresh water maker door.

The reason for the corrosion attack on 254 SMO (while titanium showed no sign of corrosion) was due to the fact that the critical temperature for crevice corrosion had been exceeded, since the service temperature was close to 60°C for the evaporators.

In an oil cooler at the Gullfaks Field the temperature for a period increased from the design temperature of 30°C to a temperature in the range of 50–60°C. This caused severe corrosion attacks in a pipe spool made from 254 SMO transporting this hot seawater. Similar failure has been reported from an onshore plant operated by Norsk Hydro at the same temperature level.

4. R & D Work in Norway

Parallel to the practical experience from the use of stainless steel in seawater systems offshore, the behaviour of stainless steel in natural seawater has been the subject of several R & D projects around the world during recent years. In Norway a programme called 'Corrosion resistant alloys for process and seawater systems' ran from 1985 to 1989. This programme, carried out by The Corrosion Centre at SINTEF, has been concerned with a number of interesting questions about the behaviour of stainless steel in seawater[3–7].

To evaluate generally how the different stainless steels resist localised corrosion the so-called Pitting Resistance Equivalent (PRE_N) has often been used. According to equation (1) this formula is based on the chemical composition of the alloy. Table 1 contains the PRE_N values for the described stainless steel alloys.

$$PRE_N = \%\,Cr + 3.3\%Mo + 16\%N \qquad (1)$$

$PRE_N > 40$ is often used as a critical limit for seawater-resistant alloys. Practical experience has shown that this is only valid under certain conditions. In the mentioned R&D project performed in Norway several parameters influence the corrosivity of stainless steels in seawater that have been examined, such as:

- The effect of biofilm formation on the electrochemical potential.
- The effect of temperature on the electrochemical potential (see Fig. 1).
- The effect of chlorination on electrochemical potential and cathodic efficiency.
- Critical temperature for localised corrosion on different welded/unwelded stainless steels.
- The influence of potential on critical temperature (see Fig. 2).

In addition, a lot of work has been done to establish a new galvanic series for different alloys in natural seawater since hitherto a number of different galvanic series have been published. However, most of these series suffer from at least two main weaknesses:

(i) They have not been measured in natural seawater;

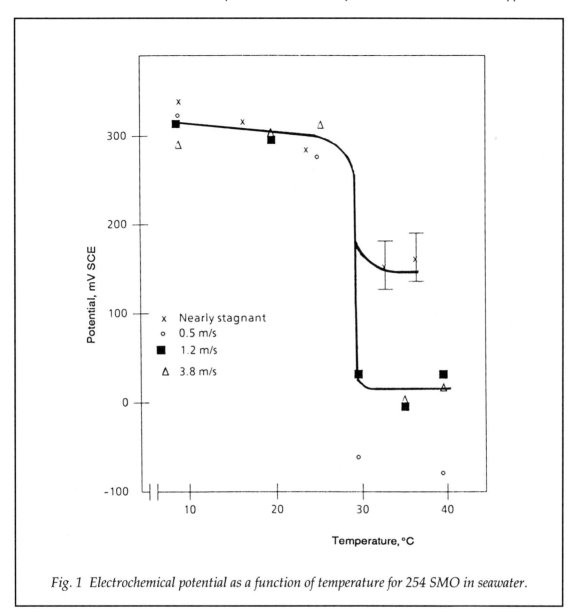

Fig. 1 Electrochemical potential as a function of temperature for 254 SMO in seawater.

(ii) The exposure period has been too short (values not stable).

Figure 3 shows the new galvanic series for a number of alloys in natural seawater at 10 and 40°C. As can be seen from the figure, the potential difference between 254 SMO and aluminium bronze (approx. equal to Ni–Al bronze) is 400–500 mV. A galvanic coupling between the two alloys will cause severe corrosion on Ni–Al bronze. This is the reason why a pump made from this alloy has to be insulated from a piping system made from 6Mo-steel (or equipped with sacrificial anodes).

As previously mentioned, a lot of new stainless steels for seawater applications have been introduced into the market over recent years. Some of these alloys are included in Table 1. To examine the corrosion properties of these alloys compared with 254 SMO, a project was initiated at Statoil R&D Centre. In this project the critical pitting temperature (CPT) of welded specimens was examined.

To define the CPT-value a potentiostatic test at a fixed potential of +400V SCE in 3% NaCl

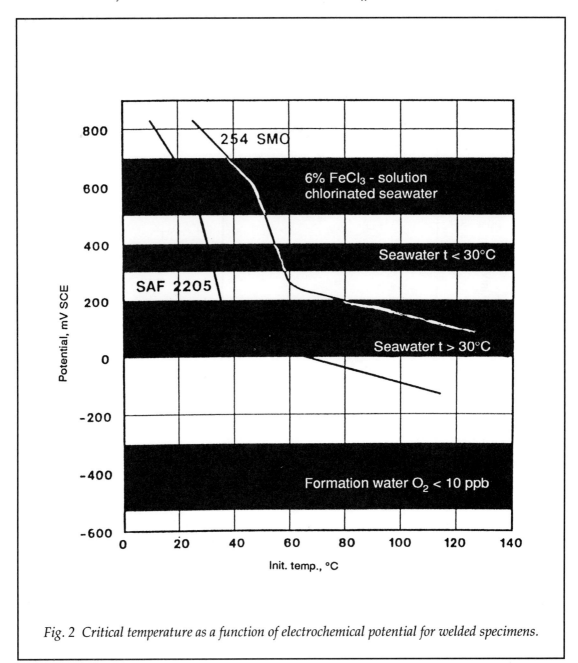

Fig. 2 Critical temperature as a function of electrochemical potential for welded specimens.

solution was used. This method compared to the more often used ASTM G-48 (modified) which has been discussed in an earlier NACE publication [7].

The test specimens were cut from circumferential welds on 4" pipes. The pipes were manually welded according to a standard specification developed by Statoil. The filler metal was Inconel 625. The specimen size was 50 × 50 mm with the weld located across the specimen. Before exposure the edges of specimens were ground to grit 500 to prevent initiation of corrosion on the edges. The specimens were not pickled before exposure. Three parallel sections of each material were exposed and each specimen was connected to the potentiostat by a platinum wire.

Each test started at 10°C and the temperature was increased in steps of 2.5°C every 24 hours. During the exposure the anodic current density was continuously recorded. CPT was defined

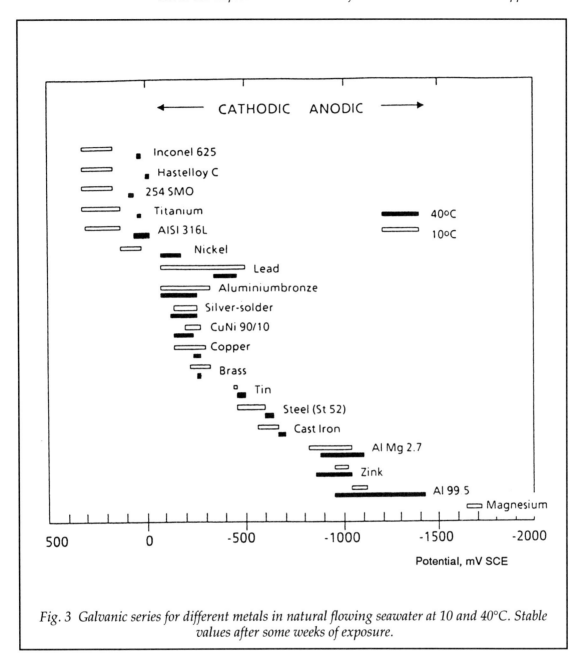

Fig. 3 Galvanic series for different metals in natural flowing seawater at 10 and 40°C. Stable values after some weeks of exposure.

as the temperature where the anodic current exceeded a current density of 10µA cm^{-2} weld area.

The results from the tests are shown in Table 4. From this table one can conclude that the steels containing 6%Mo have approximately the same critical pitting temperature in seawater. However, the anodic corrosion rate after initiation and the repassivation temperature for the different alloys have not been examined. These parameters can vary from one alloy to another.

Another important result from the tests is the high CPT-value for the duplex stainless steel SAF 2507 which equals that of the austenitic 6Mo-steels. This has been confirmed by other tests in chlorinated seawater. Based on this SAF 2507 (or equivalent 25% Cr duplex stainless steels) can be used as alternatives to 6Mo-steels in seawater systems in the future.

As an extension of this project some tests were performed to check the effect of welding

Table 4 Critical pitting temperature (CPT) of welded specimens polarized to +400 mV SCE in 3% NaCl solutions

	ALLOY				
	254 SMO	AL–6XN	Cronifer 1925 HMo	Inco 25-6Mo	SAF 2507
CPT (°C)	50 – 52.5	52.5 – 55	50 – 55	55	50

different 6Mo-alloys to each other. The CPT-values were measured according to the same procedure as described above except for the temperature, which was raised in steps of 4°C every 24 hours. Specimens were cut from circumferential welds on 3 mm thick 4" pipes. The following couples were tested:

- 254 SMO — AL-6XN
- 254 SMO — Cronifer 1925 HMo
- AL-6XN — Cronifer 1925 HMo.

The results from the tests are shown in Table 5. In most of the cases the corrosion was initiated in the heat affected zone of one of the alloys.

Comparing Table 4 and Table 5 one can see that different 6Mo-steels can be welded together without reducing the CPT-value. However, the effect on other parameters, for example mechanical properties, has not been examined.

5. Design Philosophy among Norwegian Companies

Based on the experience from service and R&D results the following design philosophy is used among Norwegian companies:

(i) AISI 316L suffers from corrosion in seawater and has to be cathodically protected to be used in this environment.

(ii) Duplex stainless steel (UNS S31803) or equivalent has to be cathodically protected to prevent corrosion attack in seawater.

Table 5 Critical pitting temperature (CPT) of welded connections polarized to +400 mV SCE in 3% NaCl solutions

	CONNECTION		
	254 SMO – AL – 6XN	254 SMO – Cronifer 1925 HMo	AL – 6XN – Cronifer 1925 HMo
CPT (°C)	48 – 56	56	48 – 52

(iii) Stainless steel with 6% Mo can be used in chlorinated seawater with a residual chlorine level in the range of 1–2 ppm at temperatures not exceeding 30°C.

(iv) Using the critical parameters described above different 6Mo-steels (according to Table 1) can replace one another.

6. Future Trends

During recent years new duplex stainless steels ($PRE_N > 40$) combining high strength and corrosion properties approximately equal to the 6Mo-steel have been developed. SAF 2507 and Zeron 100 are among these. However, in low-pressure seawater systems the high strength cannot be utilised. Based on good experience with the use of 6Mo-steel in seawater at temperatures up to 30°C combined with the experience from fabrication and welding of this material, Statoil do not see the so-called 'superduplex' stainless steel as an alternative to 6Mo-steels as piping material in seawater handling systems today—as pump or valve material 'superduplex' can be an alternative. If Statoil want to use the same alloy both for low-pressure systems (including seawater) and high pressure systems from a standardisation point of view, then the 'superduplex' is an alternative.

Besides 6Mo-steel, both titanium and glass-fibre-reinforced epoxies (GRE) are possible materials for seawater handling systems. Up to now, titanium has mostly been used in heat exchangers and for critical components. Based on a demand for reducing the maintenance costs, titanium will be an alternative to 6Mo-steels in the future, especially for temperatures exceeding 30°C.

Based on experience from installations and R&D work during the last 5–10 years, GRE will be the leading material for low-pressure piping systems containing corrosive environments— including seawater—in the future. This is based on the corrosion resistance combined with the lifetime cost of the material. Table 6 shows a comparison between costs and weight for 254 SMO and GRE for parts of the seawater lift system at Gullfaks A. According to the table the total cost relation between 254 SMO and GRE is approximately 3.3:1, while the dry weight ratio is approximately 1.4:1.

At some Statoil operated onshore and offshore plants parts of the seawater cooling systems originally made from 90/10 CuNi have been replaced by GRE because of corrosion problems.

7. Conclusions

1. 6Mo-steel (UNS S31254 or equal) has been the most widely selected material for seawater handling systems in the Norwegian sector of the North Sea during the last ten years.

Table 6 Price and weight comparison between Avesta 254 SMO and glass-fibre-reinforced epoxy (GRE) for parts of the seawater lift systems at Gullfaks A

MATERIAL	MATERIAL COSTS (£)	INSTALLATION COSTS (£)	TOTAL COSTS (£)	DRY WEIGHT (kg)
254 SMO (1984)	165.000	100.000	265.000	8470
GRE	62.000	16.000	78.000	6100

2. The experience from the use of 6Mo-steel in seawater is good.

3. Reported failures are caused by:

   - exchange of alloys (AISI 316L used instead of 6Mo);
   - use of non-recommended filler metal;
   - seawater temperatures exceeding design temperature.

4. In chlorinated seawater (< 1–2 ppm residual chlorine) 6Mo-steel can be used up to 30°C without suffering localized corrosion.

5. Different 6Mo-steels can be interchanged.

6. 6Mo-steels can be welded together without reducing the CPT-value.

7. At temperatures exceeding 30°C titanium and GRE are the most applicable materials.

8. Glass-fibre-reinforced epoxy (GRE) will replace stainless steel in low-pressure systems in the future.

9. Duplex stainless steel (UNS S31803) or lower alloyed alloys have to be cathodically protected to prevent corrosion.

References

1. R. Johnsen, Corrosion Problems in the Oil Industry, paper presented at the 11th Scandinavian Corrosion Congress, Stavanger (Norway), 1989.
2. Ø. Strandmyr, Operational Experience from the Statfjord Platforms, NITO Conference, Oslo, 26–27 October, 1988.
3. R. Johnsen and E. Bardal, The Effect of Microbiological Slime Layer on Stainless Steels in Natural Seawater, CorrosionNACE '86, Houston, Texas, paper 227.
4. R. Holte, E. Bardal and P. O. Gartland, The Time Dependence of Cathodic Properties of Stainless Steels, Titanium, Platinum and 90/10 CuNi in Seawater, Corrosion NACE '88, St. Louis, USA.
5. S. Valen *et al.*, New Galvanic Series based upon Long Duration Testing in Flowing Seawater, Paper presented at the 11th Scand. Corros. Congr., Stavanger (Norway), 1989.
6. R. Holte, The Cathodic and Anodic Properties of Stainless Steels in Seawater, thesis from NTH, Trondheim, Norway, 1988.
7. J. M. Drugli, T. Rogne, R. Johnsen and S. Olsen, Corrosion Testing of Stainless Steel Weldments in Seawater, NaCl and $FeCl_3$-Solutions, Corrosion NACE '88, NACE, St. Louis, USA.

CHLORINATION

5
Seawater Chlorination

C. MADEC, F. QUENTEL AND R. RISO

URA CNRS 322, Chimie, Electrochimie Moléculaires et Chimie Analytique, UFR Sciences et Techniques,
6, Avenue Le Gorgeu 29287 Brest, Cedex, France

Abstract

The consumption of chlorine in seawater is mainly influenced by the presence of a large excess of bromide ions and of reactive organic substances and ammonia nitrogen. It is also dependent upon physico-chemical properties such as pH. In order to account for the mechanisms which take place in natural waters, a number of experiments have been performed in synthetic media with the objective of quantifying the effect of each parameter mentioned above.

1. Introduction

Seawater and estuarine waters used as coolants in coastal power stations are almost invariably treated to prevent marine fouling with chlorine using intermittent or a low-level continuous regimes. As illustrated in Fig. 1 the effect of chlorine concentration is of great importance both in chemical and in biological corrosion.

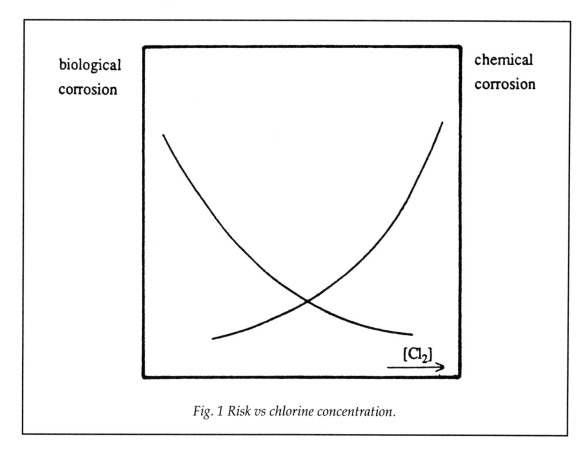

Fig. 1 Risk vs chlorine concentration.

When chlorine is added to seawater around the ppm level (1.4×10^{-5} mol l^{-1}), it undergoes a rapid consumption as it is both a powerful oxidant (E° = 1.35 V) and a chlorinating agent. A lot of inorganic and organic substances dissolved in seawater will thus react with the halogen leading to a decrease in the content of the residual oxidant with time. With the concentrations generally used and the kinetic parameters that are usually operative, the main species which can account for chlorine consumption are of three kinds:

Bromide	Ammonia	Organic substances
($10^{-5} - 8 \times 10^{-4}$ mol l^{-1})	($10^{-6} - 10^{-5}$ mol l^{-1})	(0.1 – 5 mgC l^{-1})

As the chemical composition of seawater is so complicated, the first experiments were conducted in synthetic media to quantify the effect of each species. Based on data available for the ionic strength and the acidity of seawater, the following mixture was chosen as a reference model:

$$\boxed{Na^+, Cl^- \ (0.5 \text{ mol } l^{-1}) + Na^+, HCO_3^- \ (0.03 \text{ mol } l^{-1})}$$

Bromide, ammonia and organic species were then added successively.

Only the mechanims of chlorine consumption will be studied. No details of the analytical methods used will be discussed.

2. Discussion

2.1 Bromide oxidation[1–3]

The pH of seawater and estuarine waters is usually between 6.5 and 8.5. The salinity is above 5 ‰ which means that the bromide concentration is greater than 10^{-4} mol l^{-1}. Under these conditions the thermodynamically possible oxidation of bromide by hypochlorous acid or hypochlorite is rapid:

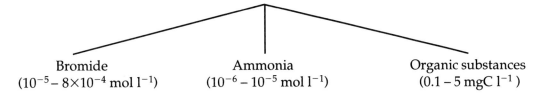

The rate law is second order; the average value of the rate contant is k = 5×10^3 l mol^{-1} s^{-1} around pH = 8 and for a salinity of 35‰. More than 99% of hypobromous and/or hypobromite is produced in less than 10 s.

The reaction is quantitative as illustrated in Fig. 2 and Table 1.

The formation of bromate as a by-product is not observed at a significant level.

As will be shown, the oxidation of Br$^-$ by chlorine is of prime importance in accounting for the main mechanisms which take place in seawater chlorination.

[It should be noted that free residual chlorine does not usually exist in seawater as bromide is in large excess; the major free species will be hypobromous acid with acidity constant ($K_a^\circ = 10^{-8.60}$).]

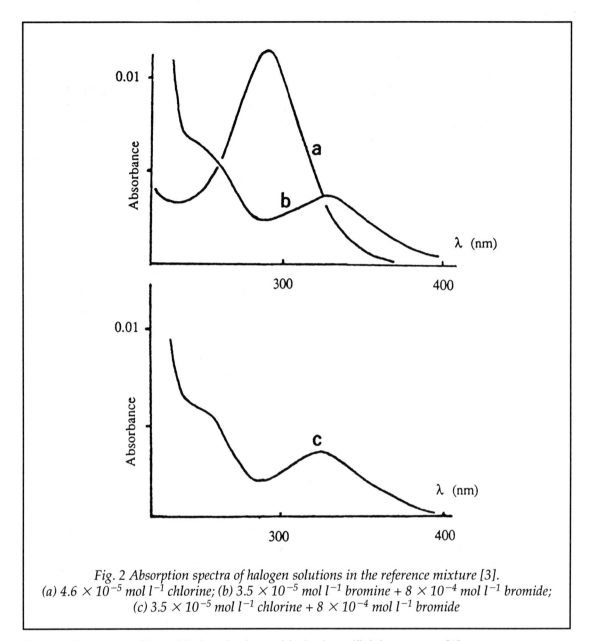

Fig. 2 Absorption spectra of halogen solutions in the reference mixture [3].
(a) 4.6×10^{-5} mol l^{-1} chlorine; (b) 3.5×10^{-5} mol l^{-1} bromine + 8×10^{-4} mol l^{-1} bromide;
(c) 3.5×10^{-5} mol l^{-1} chlorine + 8×10^{-4} mol l^{-1} bromide

Table 1 Oxidation of bromide ions by hypochlorite in artificial seawater [3]

chlorine injected mg l^{-1} Cl_2	theoretical bromine concentration mg l^{-1} Br_2	experimental bromine concentration (phenol red method) mg l^{-1} Br_2
0.85	1.92	1.96
0.56	1.28	1.26
0.28	0.64	0.61

2.2 Reactivity of ammonia in the presence of bromide [4–8]

Ammonia is known to react with chlorine and bromine forming a group of compounds called the haloamines. As ammonia nitrogen levels commonly found in coastal and estuarine seawaters are quite large and not negligible compared to the chlorine dose introduced, the mechanisms of the haloamine formation need to be investigated.

As bromide is generally found in large excess compared to the antifouling agent, and is easily oxidised by chlorine in the pH range involved, both chloramines and bromamines are able to form according to the following routes:

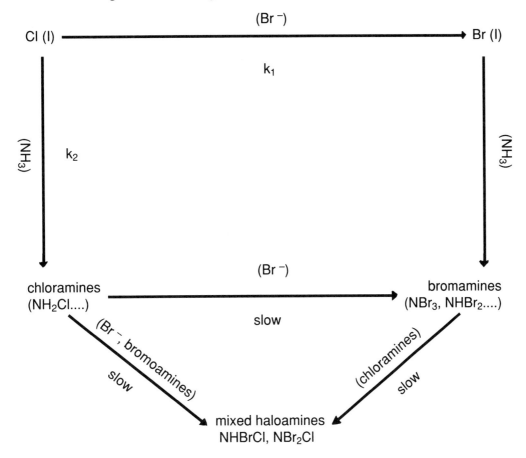

NB. Cl(I) and Br(I) mean respectively HClO and/or ClO$^-$, HBrO and/or BrO$^-$.

The k_2 rate constant (6×10^6 l mol^{-1} s^{-1}) is much higher than k_1 (3×10^3 l mol^{-1} s^{-1}); it promotes the ammonia chlorine derivative, especially the monochloramine NH$_2$Cl. However, as the concentration of bromide is much larger than that of ammonia nitrogen (NH$_4^+$ + NH$_3$), the bromide to ammonia nitrogen ratio will promote the formation of hypobromous acid and hypobromite, rather than the production of bromamines.

The predominant forms of the haloamines depend primarily on the pH and the chlorine to ammonia nitrogen ratio (r) as illustrated in the following routes (over page) and in Fig. 3.

If the r ratio is larger than 3/2, only bromine derivatives are obtained, i.e. dibromamine, tribromamine and free bromine in the +1 oxidation state. A ratio smaller than 3/2 leads to a mixture of monochloramine, dibromamine and some monobromamine; monochloramine clearly predominates when the ammonia nitrogen concentration reaches high values. Some formation of mixed haloamines is also possible.

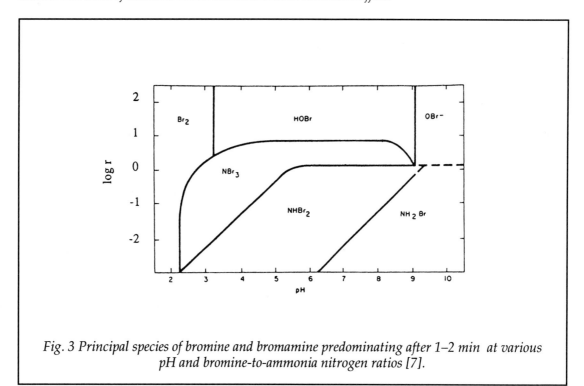

Fig. 3 Principal species of bromine and bromamine predominating after 1–2 min at various pH and bromine-to-ammonia nitrogen ratios [7].

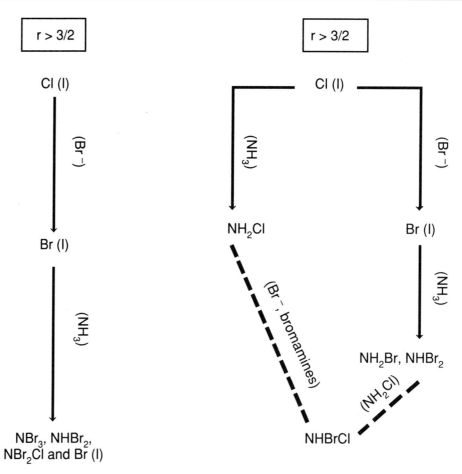

The 3/2 value results from the reaction stoichiometry: $2NH_3 + 3HBrO \rightarrow N_2 + 3H_3O^+ + 3Br^-$, and corresponds to the breakpoint mole ratio which is due to the preferential formation and rapid decomposition of dibromamine.

Bromamines and bromochloramines are less stable than chloramines. Once they have formed, they undergo hydrolysis decomposition as follows:

$$2NBr_3 + 6H_2O \rightarrow N_2 + 3HOBr + 3H_3O^+ + 3Br^-$$

The main parameter which governs the decomposition rate is the pH, as reported in Fig. 4.

2.3 Reactivity of organic substances in the presence of bromide [9–16]

Evidence that dissolved organic matter is the most significant source of chlorine demand of seawater used in cooling circuits is now well recognized [9]. Nevertheless the specific behaviour of the different kinds of compounds which characterise the organic material has seldom been studied and has been even less often quantified. This is not surprising as less than 20% of the organic substances have so far been identified.

Our experiments have been carried out in two ways with the objective of providing a common approach to the problem, and a detailed investigation of the two classes of compounds occurring in biological cycles (amino and fatty acids).

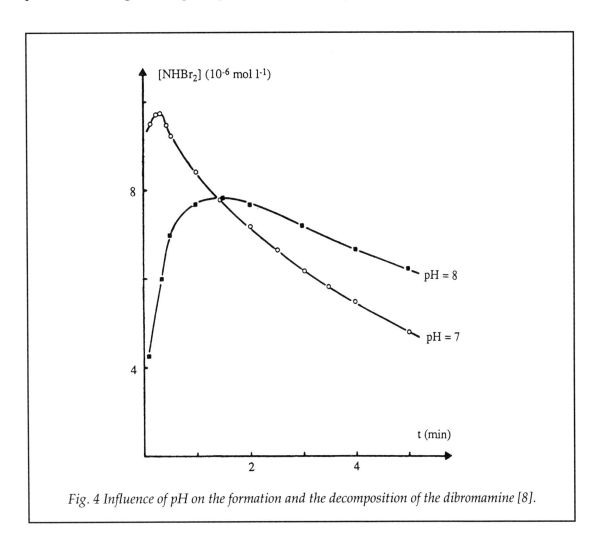

Fig. 4 *Influence of pH on the formation and the decomposition of the dibromamine [8].*

2.3.1 Humic substances [10–15]

The experiments which refer to the common approach (i) above have been carried out with humic and fulvic acids as model substrates of the unknown fraction. These materials were extracted from aqueous samples according to Malcolm's procedure [10].

As illustrated in Fig. 5, chlorine undergoes a rapid consumption in two distinct phases when it is added at the ppm level to an aqueous sample of fulvic substances: a fast initial demand is followed by a slower phase [14]. These substances, which are a mixture of complex compounds, have a lot of reactive sites such as aromatic rings activated by side functional groups. For example, one molecule of 1,3 dihydroxybenzene consumes five molecules of injected chlorine; however, such a compound is probably less reactive when it is combined in a polymeric structure.

As illustrated in Fig. 6, the consumption of chlorine by organic matter increases and the reaction kinetics are faster when bromide ions are present, as bromine is a better halogenating agent than chlorine. However, according to Helz [15] a large fraction of the antifouling agent is consumed in oxidizing reactions; more than 30% would be used to produce carbon dioxide. The important effect of Br^- on the reactivity of chlorine towards the organic matter is also shown by the percentage of haloform by-products vs salinity. The main compound which forms is $CHBr_3$, in agreement with the previous data.

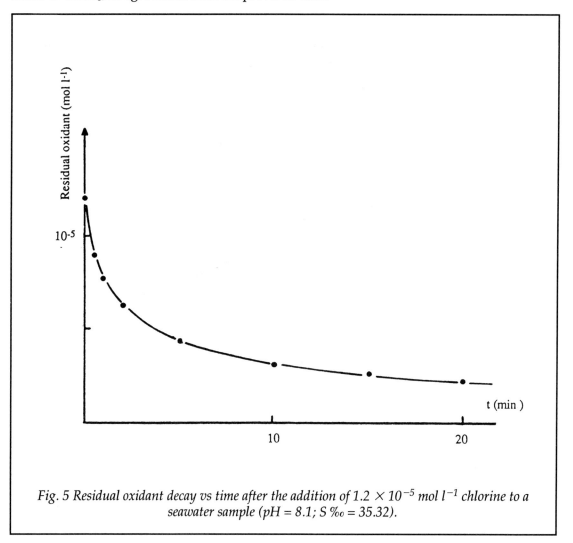

Fig. 5 Residual oxidant decay vs time after the addition of 1.2×10^{-5} mol l^{-1} chlorine to a seawater sample (pH = 8.1; S ‰ = 35.32).

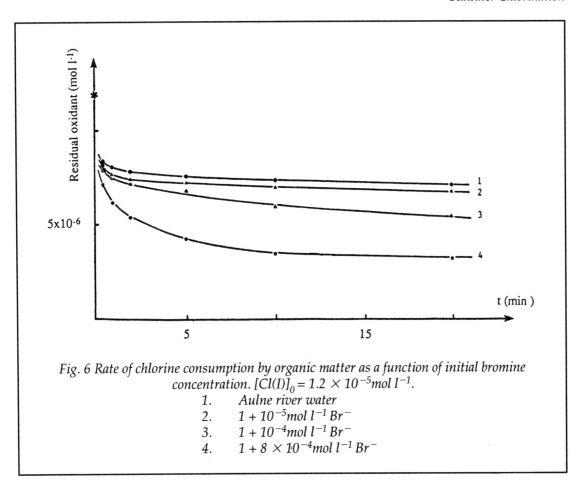

Fig. 6 *Rate of chlorine consumption by organic matter as a function of initial bromine concentration.* $[Cl(I)]_0 = 1.2 \times 10^{-5}$ mol l^{-1}.
1. Aulne river water
2. $1 + 10^{-5}$ mol l^{-1} Br^-
3. $1 + 10^{-4}$ mol l^{-1} Br^-
4. $1 + 8 \times 10^{-4}$ mol l^{-1} Br^-

2.3.2 Amino acids and fatty acids [16 and 17]

Chlorine reacts rapidly with a number of organic substances particularly when bromide ions are present in excess. Species which are essential for the growth of zoo- and phytoplankton can thus be degraded. Our objective was therefore to investigate how amino acids and fatty acids—two classes of compounds occurring in biological cycles—behave towards the halogen under the two main chlorination regimes.

2.3.2.1 Amino acids [16]

The amino acids dissolved in seawater are free or combined in polymeric structures like proteins. In the case of the free compounds our data show a fast consumption of the halogen by the α amino acids while the oxidative power of the sample is only slightly affected by aliphatic amines or β and γ amino acids, of the same length (Fig. 7). However, this does not mean that these compounds are unreactive towards the halogen; on the contrary, the by-products which form are stable and have oxidative properties.

Only the reactivity of the α compounds was studied afterwards as they represent the largest fraction of the free amino acids encountered in natural waters.

Hypohalites (HOX, XO$^-$) react with α amino acids to form a N-monohalamine or a N, N'-dihalamine depending on the initial halogen/amino acid ratio and on the acidity of the sample. These halo-compounds are highly unstable and decompose rapidly to produce CO_2, NH_3 and the corresponding aldehyde or nitrile as follows:

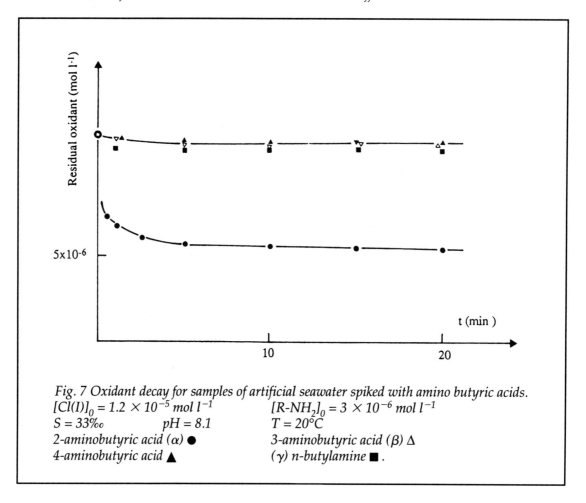

Fig. 7 Oxidant decay for samples of artificial seawater spiked with amino butyric acids.
$[Cl(I)]_0 = 1.2 \times 10^{-5}$ mol l^{-1} $[R-NH_2]_0 = 3 \times 10^{-6}$ mol l^{-1}
$S = 33‰$ $pH = 8.1$ $T = 20°C$
2-aminobutyric acid (α) ●   3-aminobutyric acid (β) △
4-aminobutyric acid ▲   (γ) n-butylamine ■ .

$$X_2N\text{-}CH\,COO^- + H_2O \rightarrow N\equiv C\text{-}R + 2\,X^- + HCO_3^- + 2\,H^+$$
$$\quad\quad\;\;|$$
$$\quad\quad\;\;R$$

$$XHN\text{-}CH\text{-}COO^- + 2\,H_2O \rightarrow O=C\text{-}H + HCO_3^- + NH_4^+ + X^-$$
$$\quad\quad\;\;|\quad\quad\quad\quad\quad\quad\quad\;\;|$$
$$\quad\quad\;\;R\quad\quad\quad\quad\quad\quad\quad\;R$$

These decarboxylation mechanisms do not explain the consumption yield that is observed with a few compounds. The differences between some of them are associated with the role of the side group R on the kinetics and the efficiency of the consumption of the oxidant (Fig. 8).

When α amino acids are bound in combined structures, the data are different. The amide nitrogen bond is resistant to aqueous HOX at room temperature, and these combined compounds react with the halogen only through the -NH$_2$ terminal function in the same way as long chain aliphatic amines.

In natural waters, the measurement of the residual oxidant only does not allow the impact of chlorination towards amino acids to be quantified since many organic compounds react with the halogen. A suitable HPLC method was developed. A few experiments carried out on samples of seawater from the 'Rade de Brest' treated with 1 ppm of chlorine have shown that around 5% of the halogen dissipated during the first 3 min is consumed by the dissolved free amino acids, the corresponding depletion of which is about 50% (Table 2).

Table 2 Depletion (%) of free α amino acids after chlorination (1.2×10^{-5} mol.l^{-1}) of an estuarian water sample

S = 33.62 ‰   pH = 8.07   T = 20°C

compound \ time (min)	1.5	10	50
alanine	49 ± 5	45 ± 5	43 ± 5
glycine	28 ± 5	51 ± 5	68 ± 5

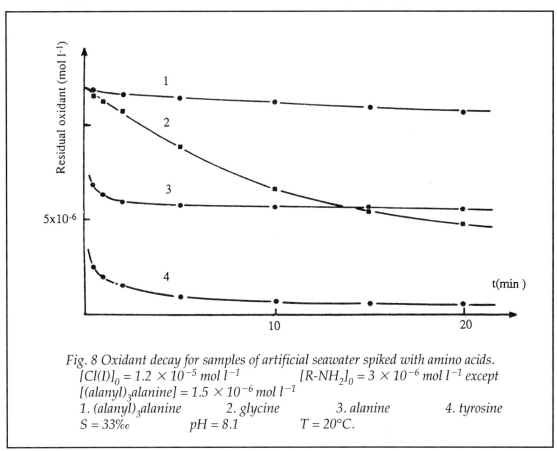

Fig. 8 Oxidant decay for samples of artificial seawater spiked with amino acids.
$[Cl(I)]_0 = 1.2 \times 10^{-5}$ mol l^{-1} $[R-NH_2]_0 = 3 \times 10^{-6}$ mol l^{-1} except
$[(alanyl)_3 alanine] = 1.5 \times 10^{-6}$ mol l^{-1}
1. (alanyl)$_3$alanine   2. glycine   3. alanine   4. tyrosine
S = 33‰   pH = 8.1   T = 20°C.

2.3.2.2 Fatty acids [17]

As before, experiments were carried out at first in synthetic seawater samples spiked with a few aliphatic fatty acids and methyl esters (*ca*. 10^{-8} mol l^{-1} each). The chromatograms depicted in Fig. 9 illustrate (i) the lack of reactivity of the saturated compounds; this means that there is no oxidative decarboxylation as in the case of the α amino acids; and (ii) the depletion of the unsaturated species; the reactions which take place are more complex than a simple attack of the double-bond (addition, oxidation).

HPLC measurements carried out in more concentrated mixtures (10^{-5} mol l^{-1}) show, to a first approximation, that addition reactions are always minor pathways for the acidic levels usually encountered i.e. 6.5 < pH < 8.5, even in the presence of bromide ions. Experiments performed with oleic acid reveal that a pH increase promotes this route.

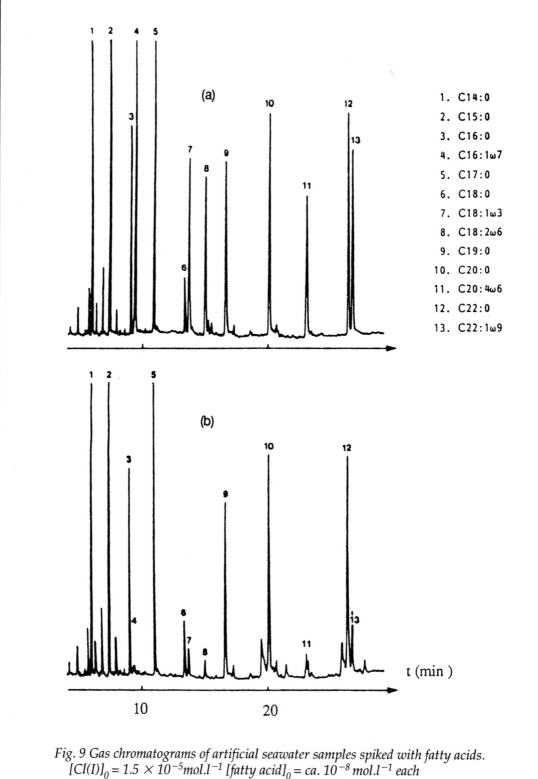

Fig. 9 Gas chromatograms of artificial seawater samples spiked with fatty acids.
$[Cl(I)]_0 = 1.5 \times 10^{-5} mol.l^{-1}$ $[fatty\ acid]_0 = ca.\ 10^{-8}\ mol.l^{-1}$ each
$S = 33‰$ $pH = 8.1$ $T = 20°C$
A unchlorinated sample **B** chlorinated sample (3 min)
saturated compounds: 1, 2, 3, 5, 6, 9, 10, 12
unsaturated compounds: 4., 7, 8, 11, 13.

The main reactions would thus involve the cleavage of the double-bonds of the unsaturated compounds. As no by-product has yet been identified, we cannot propose an accurate degradation mechanism as for α amino acids. Under experimental conditions which allow the oxidant decay to be measured (2 < ratio halogen / acid or ester < 5), we find in the case of oleic acid, a single double-bond species, a consumption of one molecule of chlorine (or bromine) per molecule of solute. Nevertheless, when this ratio is greater than 1000, which is often the case under true chlorination conditions, it is not out of the question that the consumption will be larger; the absence of mono- or diacid C_8-C_9 species as by-products in the case of oleic acid means that complex reactions would take place.

The main parameters which are responsible for the kinetics and the variations in efficiency variations are of three types:

(i) Typical solute parameters, i.e. nature of the compound (acid or ester) i.e. length of the aliphatic chain; number of double bonds

(ii) Sample parameters, i.e. acidity–concentration of bromide ions; presence of inhibiting substances (ammonia nitrogen, reactive organic species).

(iii) Chlorination conditions. Chlorination of natural seawater is usually conducted in two different ways: i.e. by a continuous halogen injection at a low level (0.1 ppm), or by an intermittent halogen addition at a high level (> 1 ppm).

The stability of fatty acids and esters was investigated in each case.

When the oxidant is injected in excess (free halogen can still be measured after a few minutes) the reactions which take place are to a first approximation similar to those observed in synthetic samples. All unsaturated lipids (acids and esters) are more or less degraded depending on the parameters stated above (Table 3).

Table 3 Effect of solute parameters on the depletion of fatty acids (FA) and fatty esters (FE) in natural seawater

$[Cl(I)]_0 = 1.2 \times 10^{-5}$ mol l^{-1} $[FA]$ or $[FE] = 2 \times 10^{-8}$ mol l^{-1}

S = 34.48 ‰   pH = 8.12   3 min contact time

compound	palmitoleic acid (C_{16}:1ω7)	erucic acid (C_{22}:1ω9)	methyl palmitoleate (C_{16}:1ω7)
depletion (‰)	70 ± 7	25 ± 7	40 ± 7

If chlorine is added at a low level only short-chain unsaturated acids partly react. Under these chlorination conditions the presence of more reactive organic species, such as humic substances, play a prominent part.

3. Conclusion

When chlorine is injected into seawater as antifouling of bactericidal agent, it is important to be able to measure the residual oxidant and to know the fate of the consumed fraction. Many analytical methods provide a quantitive determination of the free and combined residual oxidant. However, characterising all the species which arise from such treatments is not possible since in many of them the few hundred organic compounds that are produced are at a very low level.

References

1. L. Farkas, M. Lewin and R. Bloch, J. Am. Chem. Soc., 1949, **71**, 1988.
2. G. T. F. Wong and J. A. Davidson, Water Res., 1977, **11**, 971.
3. A. Peron and J. Courtot-Coupez, Water Res., 1980, **14**, 329.
4. G. W. Inman, T. F. Lapointe and J .D. Johnson, Inorg. Chem., 1976, **15**, 3037.
5. G. W. Inman and J. D. Johnson, 'Water Chlorination, Environmental and Health Effects', Vol. 2, Ed., Ann Arbor, 1978.
6. A. Peron and J. Courtot-Coupez, Water Res., 1980, **14**, 883.
7. J. D. Johnson, 'Disinfection Water and Wastewater', Ed. Ann Arbor, 1975.
8. J. L. Cromer, G. W. Inman and J. D. Johnson, 'Chemistry of Waste water Technology', Vol. 2, Ed, Ann Arbor, 1978.
9. G. T. F. Wong and T. J. Watts, Water Res.,1984, **18**, 501.
10. E. M. Thurman and R. L. Malcom, Environ. Sci. Technol., 1981, **15**, 463.
11. A. C. Sigleo, G. R. Helz and W. H. Zoller, Environ. Sci. Technol., 1980, **14**, 673.
12. R. G. Qualls and J. D. Johnson, Environ. Sci. Technol., 1983, **17**, 692.
13. T. Ishikawa, T. Sato, Y. Ose and H. Nagase, Sci. Total Environ., 1986, **54**, 185.
14. J. C. Goldman, H. L. Quinby and J. M. Capuzzo, Water Res., 1979, **13**, 315.
15. G. R. Helz, D. A. Dotson and A. C. Sigleo, 'Water Chlorination, Environmental and Health Effects', Vol. 2, Ed., Ann Arbor, 1978.
16. C. Madec, B. Trebern and J. Courtot-Coupez, Water Res., 1985, **19**, 1171.
17. B. Trebern, Y. Marty, F. Quentel and C. Madec, Environ. Technol. Letters, 1987, **8**, 235.

6
Experiences with Seawater Chlorination on Copper Alloys and Stainless Steels

P. GALLAGHER*, A. NIEUWHOF AND R. J. M. TAUSK

Shell Research, Arnhem, The Netherlands
Experimental work has been largely carried out by Thornton Research Centre, UK
*Currently with Thornton Research Centre, PO Box 1, Chester, UK

1. Introduction

The seawater handling systems on offshore platforms serve a variety of tasks. They supply cooling water to such equipment as crude oil and gas coolers, service water for drilling operations, downhole injection and storage ballast, as well as supplying water for fire mains systems and living quarters. To control fouling such systems are routinely chlorinated. In this paper we present a brief description of chlorination practice and materials performance over several years in existing seawater handling systems (both offshore and shipboard), which are in the main constructed from copper–nickel alloys. In addition, some aspects of our research into the use of corrosion resistant alloys as alternative materials for the fabrication of these systems is presented together with the effect that chlorination could have on the corrosion behaviour of these newer alloys.

2. Existing Systems and Practices

Much of the design and operation philosophies for seawater handling systems for early offshore platforms has been based on 'tried and tested' designs from the shipboard Shell Fleet operations. These consisted essentially of careful screening of sea inlets to prevent larger objects being drawn into systems and the injection of hypochlorite as a biocide to produce a hostile environment to either discourage fouling settlement and growth or remove existing fouling (Fig. 1).

2.1 Chlorination practice

During the 1970s electrochlorination units were fitted into most vessels in the Shell Fleets to replace the need for storage and injection equipment for hypochlorite. Various problems were reported [1] when operating the electrochlorination units, including platinum loss from electrodes, cell blockage by calcareous deposits, breakdown of electrical insulation between the cell and pipework, in addition to 'mechanical' problems such as flow switch malfunctions and cell and pipework leakage. Investigation into the causes of platinum loss showed that this occurred at high cell overpotentials, resulting from lower conductivity of the seawater either at low temperatures (< 10°C) or when operated in low salinity waters (estuarine locations). Limiting of the cell voltage at the electrochlorinator power supply eliminated this problem. Several instances of anode insulation failure were reported resulting in catastrophic corrosion of adjacent titanium valves and pipework (Fig. 2).

2.2 Levels of chlorination

For the copper alloy based seawater handling systems on vessels a level of free chlorine of 0.5 ppm at seawater feed and 0.1 ppm at main condenser outlets was normally employed. This

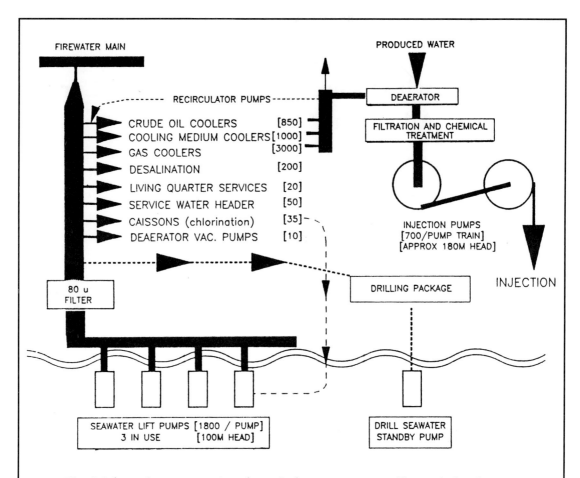

Fig. 1 Schematic representation of a typical seawater system (figures in brackets are typical flow rates, $m_3 \, h^{-1}$).

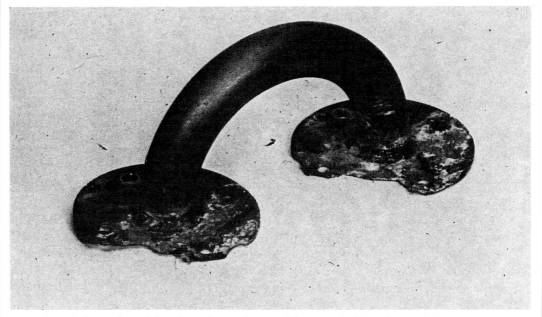

Fig. 2 Corrosion of titanium pipework caused by electrical insulation failure within the elctrochlorinator system.

level was found to be adequate to control fouling to an acceptable level in most situations. During deep-sea operation of the vessels such levels of chlorination were sometimes limited to 4 hours per day as an intermittent dosing. This was in part to allow an intermittent dosing of the system with ferrous sulphate which was used to promote the formation of a hard, erosion-resistant corrosion product on the condenser tubes. The formation of this harder layer did not occur as readily in the presence of chlorine. A disadvantage found with intermittent chlorination was that certain fouling appeared to become tolerant to the treatment, necessitating the use of significantly higher levels of chlorination (several ppm) during the dosing periods.

Higher levels of chlorine were found to be necessary for treating the seawater on offshore platforms to achieve the desired control of fouling even when using continuous injection. This is a consequence both of the longer residence time of water in the systems and the storage of water in the storage cells for relatively long periods (days, weeks) which increases the biological demand. Thus, a residual chlorine level of between 0.8 and 1.0 ppm is frequently used as a target level, as measured at the main seawater header.

The materials used for offshore seawater handling systems have, at various periods, differed from the traditional copper alloys found shipboard. Early platform systems used a large amount of concrete lined steel pipework, which had a very mixed success, dependent essentially on how good the concrete lining was. The effect of chlorination did not seem to feature at this stage. Later systems used more copper–nickel alloys for the aerated sections, but for fire mains (where aeration was intermittent), carbon steel was still employed. For small diameter pipework where flow velocities in excess of the limits allowed for copper–nickel (> 3.5 ms^{-1}) were envisaged, Monel 400 was substituted for the copper–nickel. A noticeable effect of chlorination on Monel 400 when welded with a matching filler, Monel 60, had been noted and research work carried out to assess the effect. A brief review of this work is presented later in this paper.

3. Use of the Newer Corrosion Resistant Alloys (CRAs) for Seawater Systems

Considerable interest has been shown in recent years in the possibilities of using the newer corrosion resistant alloys of the 6 % Mo austenitic type and the 25Cr-5Ni group of duplex stainless steels. The main advantage of using such alloys for the construction of offshore platform seawater systems is in the ability to construct smaller diameter, thinner walled and thus lighter weight systems by utilising significantly higher water velocities (7 ms^{-1} compared with 3 ms^{-1} for Cu-based alloys).

Development of pilot seawater handling systems and assessment of the properties of these stainless steels under realistic operating conditions, including the use of electrochlorination, has been carried out. Some of the results of this work pertaining to chlorination effects are presented later in this paper.

4. Experimental Work

The experimental work described in this paper was carried out in natural flowing seawater at our marine test station in Holyhead, N. Wales. When required, the seawater was chlorinated using an electrochlorination system (Chloropac, Electrocatalytic), and the residual chlorine level measured by a colorimetric method using DPD1 reagent and a Lovibond comparator. To minimise fluctuations in the level of chlorine generated, which would arise from flow changes in the system, a signal from electromagnetic flowmeters (Mag-X, Fisher and Porter) was used to modulate the electrochlorinator current.

5. Chlorination Effects Observed on Monel 400

Corrosion damage was reported from offshore operations of small sections of seawater cooling systems constructed in the nickel–copper alloy, Monel 400. The damage was predominantly in the form of preferential attack to each side of the welds, which had been made using the recommended matching filler Monel 60, but was also found in the form of pitting. Similar Monel systems had been used previously at other locations without any reported problems. The only apparent difference was the uncertainty of chlorination practices in this instance. An example of the type of attack is shown in Fig. 3.

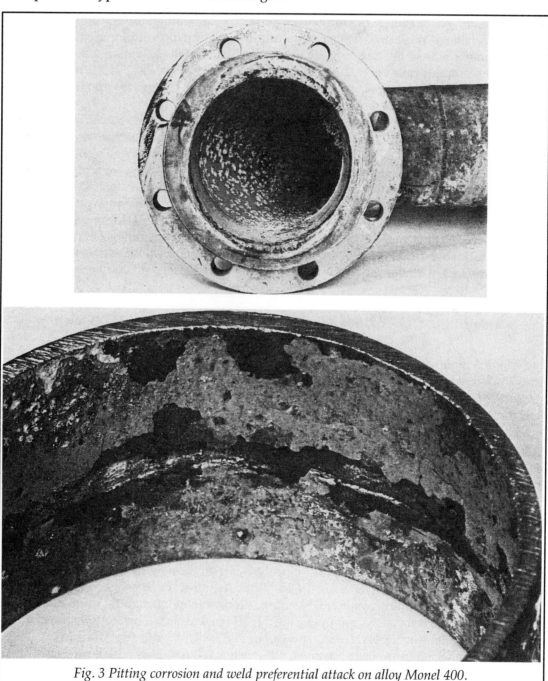

Fig. 3 Pitting corrosion and weld preferential attack on alloy Monel 400.

Exposure testing of welded pipe sections at the marine test station highlighted an extreme difference in corrosion damage between samples exposed to unchlorinated seawater and those exposed to seawater which was continuously chlorinated to 0.8ppm. Figure 4 shows an example of samples subjected to these two test conditions after 6 months exposure. Monitoring of the corrosion potential of these samples had been carried out during the exposure period, the results of which are shown in Fig. 5.

Fig. 4 Monel 400 samples exposed to natural and chlorinated seawater for 6 months. (a) Unchlorinated seawater exposure; (b) Exposure to seawater chlorinated to 0.8 ppm.

Studies of the potentiodynamic behaviour of the Monel 400 alloy had been carried out concurrently. The anodic curve, shown in Fig. 6, indicated that a passive/active transition occurred in the region of 0 mV SCE. This was further confirmed by potentiostatic testing at three levels –50, 0 and 50 mV (Fig. 7), from which it was clearly evident that between –50 and 0 mV the passive response of the alloy was lost. Comparison of these data with the potentials measured from exposure tests showed that in the unchlorinated seawater the alloy was in an active state, whereas in seawater chlorinated to 0.8 ppm, lower potentials were measured, which relate to a passive region for this alloy in the seawater environment.

The reason for the lower potentials measured in unchlorinated seawater is not proven, however, it is well recognised that the formation of a biolayer on the surface of metals exposed to natural seawater assists the oxygen reduction reaction, thus promoting higher potentials. At low levels of chlorination this biofilm effect is lost and the oxidising nature of the chlorine has insufficient effect to raise the corrosion potential. At higher chlorination levels it is envisaged that once again higher corrosion potentials may be seen with consequences similar to that of unchlorinated water.

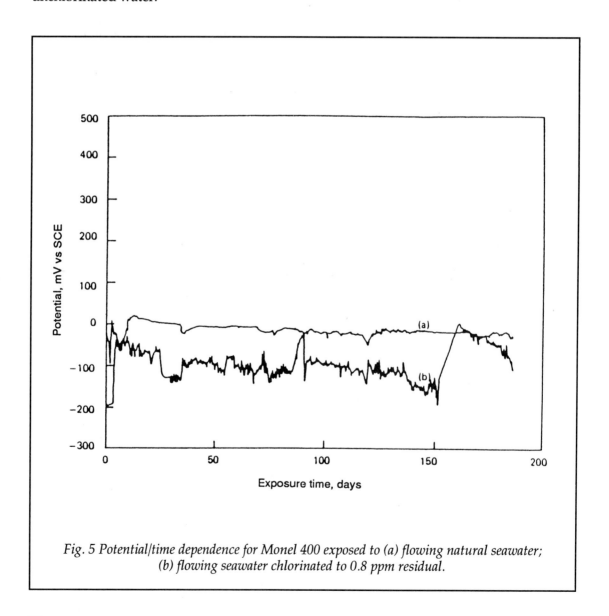

Fig. 5 Potential/time dependence for Monel 400 exposed to (a) flowing natural seawater; (b) flowing seawater chlorinated to 0.8 ppm residual.

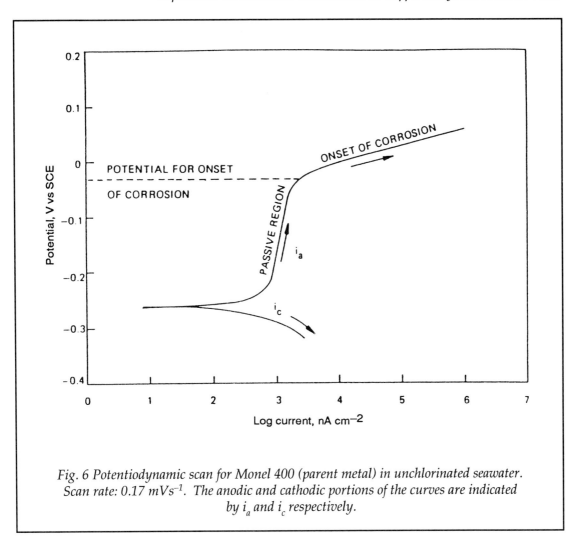

Fig. 6 Potentiodynamic scan for Monel 400 (parent metal) in unchlorinated seawater. Scan rate: 0.17 mVs^{-1}. The anodic and cathodic portions of the curves are indicated by i_a and i_c respectively.

6. Chlorination Effects on CRAs

6.1 Ambient temperature testing

Initial work concentrated on studying the effect of chlorine on the electrochemical potential attained by 6% molybdenum-containing stainless steel and duplex alloys of the 25Cr-5Ni type on exposure to seawater. This study was important since previous work [2] had demonstrated that the initiation of crevice corrosion on stainless steel alloys in seawater was strongly dependent on the potential attained by the alloy, with more positive potentials resulting in an increased probability that corrosion will occur.

In the first series of tests, 150 × 100 mm plate samples of the CRAs were exposed in tanks to seawater flowing at a nominal 0.1 ms^{-1} and with residual chlorine levels between 0 and 0.8 ppm. One half of the samples had a polyacetal crevice washer (Fig. 8) clamped upon them to produce numerous possible crevice sites, the other half of the samples were exposed as bare plates. The tests extended over a period of approximately 6 months, during which time the seawater temperature varied between 10 and 16°C. For the whole of the exposure period the potentials of the samples, referenced to a saturated calomel electrode, were measured on a 6 hourly basis using a Hewlett Packard 3497A data acquisition unit controlled by a HP85B microcomputer.

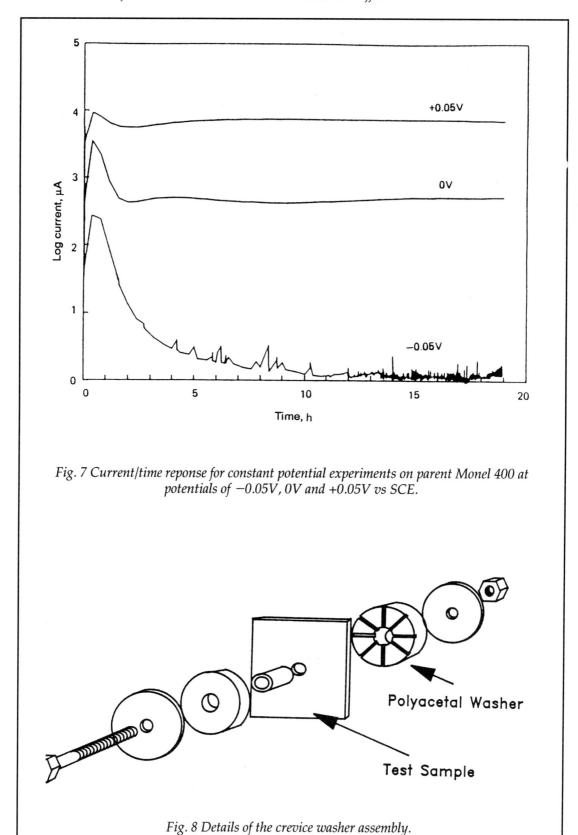

Fig. 7 Current/time reponse for constant potential experiments on parent Monel 400 at potentials of −0.05V, 0V and +0.05V vs SCE.

Fig. 8 Details of the crevice washer assembly.

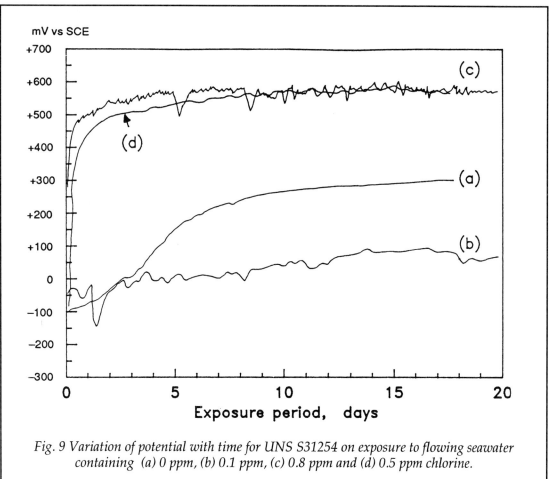

Fig. 9 Variation of potential with time for UNS S31254 on exposure to flowing seawater containing (a) 0 ppm, (b) 0.1 ppm, (c) 0.8 ppm and (d) 0.5 ppm chlorine.

The results obtained, over the first 20 days exposure, for a 6% Mo alloy plates to standard UN S 31254 are depicted in Fig. 9. Very similar results were obtained for other 6% Mo CRAs and also for the 25Cr-5Ni duplex alloys, whether or not a crevice washer was present. As can be seen from this figure, after about 10 to 15 days the samples reached a reasonably steady electrochemical potential. Figure 10 shows a plot of the steady state potentials reached at the different chlorination levels.

The rise in potential over the first 15 days of exposure to unchlorinated seawater is a widely observed phenomenon on many stainless steel alloys [3, 4]. The effect has been attributed to catalysis of the cathodic reduction reaction on the steel surface (oxygen reduction) by biofilms of marine organisms and bacteria, a theory which is supported by the results shown in Fig. 10, where chlorination of the seawater to a level just sufficient to kill the marine organisms, leaving a barely detectable residual level, prevented the potential rise.

The stable potential attained by the alloy (*ca.* 50 mV) in the low residual chlorine seawater was very similar to that seen in artificial seawater, where it has been shown to result in a considerable reduction in localised corrosion of alloys normally susceptible to crevice corrosion (e.g. AISI 316L) by reducing the chances of pit initiation. A later exposure test on the AISI 904L alloy shows that the low potentials in the low residual chlorine seawater also favour a low corrosion rate (Fig. 11). In this instance the difference between the corrosion on samples exposed to low residual chlorine levels (where no initiation of corrosion occurred) and unchlorinated seawater is quite marked. Low chlorination levels, therefore, appear to have a

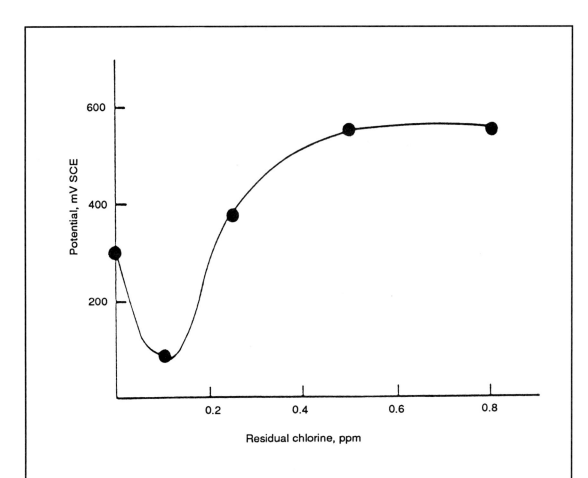

Fig. 10 Dependence of the stable potential attained by UNS S31254 on residual chlorine level.

Fig. 11 Crevice washer plates for alloy AISI 904L exposed for 50 days to (a) unchlorinated seawater, and (b) seawater chlorinated to < 0.1 ppm residual.

beneficial effect on the corrosion of CRAs by maintaining low potentials, thus reducing the probability of initiating localised corrosion.

6.2 Dependence of potential on flow rate

Potential/flowrate/chlorine level relationships were determined in a large-scale pipe test facility. This consisted of a 30 m length of 50 mm bore pipe in alloy UNS S31254, flanged at 6 m intervals, through which a flowrate in excess of 20 ms^{-1} of once-through seawater could be maintained at residual chlorine levels up to 1.5 ppm.

At each of the chlorination levels studied, the system was allowed to stabilise for a week at a flow rate of 10 ms^{-1}. Following this stabilising period the potential of the pipe section was continuously monitored using a high impedance voltmeter with the output linked to a chart recorder. The flowrate was adjusted stepwise from 0.5 to 15 ms^{-1} during a period of approximately 2 h, allowing a steady state to be reached at each flowrate. Additionally, this pipeline was exposed for a 7 month period to seawater flowing at 10 ms^{-1} and chlorinated to residual levels of 0.4 ppm initially (for 75 days), followed by 0.8 ppm for the remainder of the test.

The results of this series of tests are shown in Fig. 12. At low residual chlorine levels and without chlorination (Fig. 12(a) and (b)), the potential increased by only 100 mV as the flow was increased from 0.5 to 15 ms^{-1}. At higher residual chlorine levels, however, the potential increased rapidly with flow rate from values of around 400 mV SCE at 0.5 ms^{-1} to 650 mV SCE at 15 ms^{-1}, with the actual relationship being dependent on the chlorine level, as shown in Fig. 12(c) and (d).

The potential of the pipe over a 7 month operating period at a seawater flow rate of 10 ms^{-1} is shown in Fig. 13. The potentials observed at the two chlorination levels fluctuated with time but were generally in agreement with the values expected from the short term tests (Fig. 12). Even at these high potentials no localised corrosion was apparent on the UNS S31254 pipes or flanges after the 7 month exposure period.

6.3 Testing in heated seawater

Having established that under normal chlorination conditions (0.8 ppm nominal) and ambient temperatures that the 6 % Mo austenitic and 25Cr-5Ni stainless steels were resistant to localised corrosion damage in seawater, the work programme continued by examining the combined effects of elevated seawater temperatures, as would be expected downstream of coolers, and higher chlorination levels as is often found in seawater systems near to hypochlorite/chlorinated water injection points. Three samples of three CRAs (two 6 % Mo austenitic and one duplex stainless steel; compositions as Table 1) were exposed as crevice washer samples in each of three tanks, which were controlled at 70, 40 and 30 °C, for a period of 100 days. During this period the potential of each sample was monitored. Three series of tests were carried out in which the chlorination level through the three tanks was controlled at zero, 1.3 and 3.0 ppm. A schematic diagram of the test rig utilised is shown in Fig. 14.

While testing the crevice washer plates two distinct areas of localised corrosion were evident under the various test conditions. These areas are indicated in Fig. 15 and described below.

The first was the intended crevice area formed between the polyacetal washer and the test plate below the water surface. The second common site for localised attack was beneath salt deposits which formed on the plates above the water-line. This production of a salt layer was predominantly at 40 °C. At 70 °C the salt deposits tended to be 'washed back' into the tank by the high levels of condensation. Details of the extent of corrosion attack on the samples of stainless steels, for the three temperature conditions 30, 40 and 70 °C and for the various chlorination levels are presented in Fig. 16.

Examination of this figure shows that in the fully submerged crevice washer area all of the alloys were unattacked at all three temperatures in seawater containing zero or 1.3 ppm

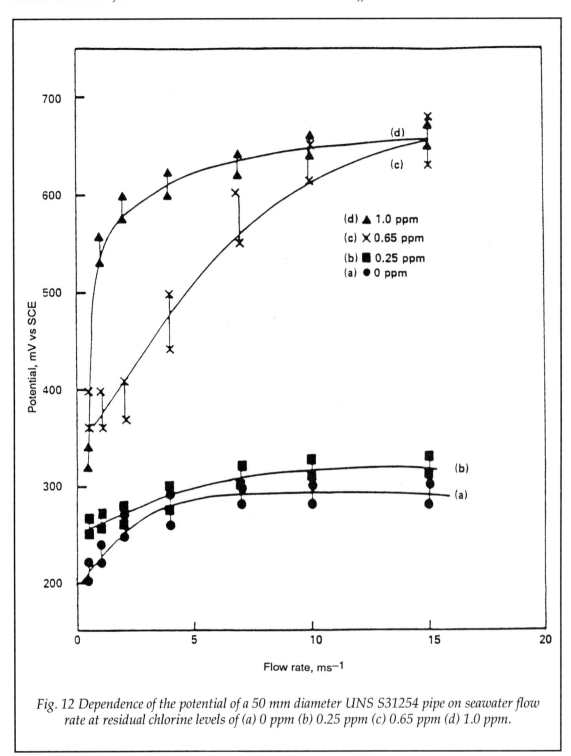

Fig. 12 Dependence of the potential of a 50 mm diameter UNS S31254 pipe on seawater flow rate at residual chlorine levels of (a) 0 ppm (b) 0.25 ppm (c) 0.65 ppm (d) 1.0 ppm.

chlorine. This compares with the results at 3 ppm in which significant localised attack occurred. However, none of the alloy/temperature combinations proved to be completely immune to all forms of localised corrosion attack.

Details of the corrosion damage in the seawater chlorinated to 3 ppm are presented in Fig. 17. For all three alloys the number of pit sites increased with temperature. This supports the

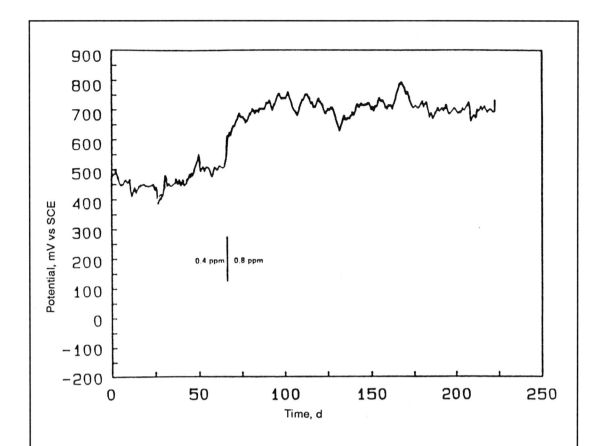

Fig. 13 Potentials measured of a UNS S31254 tube at chlorination levels of (a) 0.4 ppm for the first 75 days and (b) 0.8 ppm thereafter; both at a flow rate of 10 ms^{-1}.

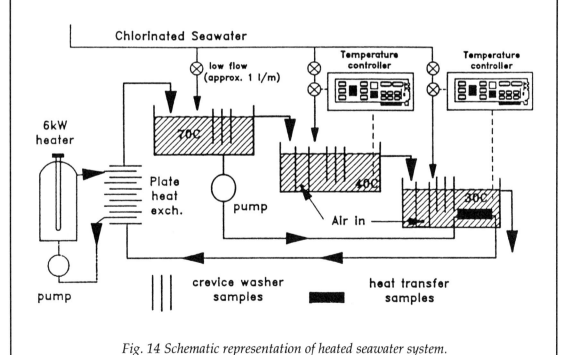

Fig. 14 Schematic representation of heated seawater system.

Table 1 Chemical analysis of alloys tested in heated seawater

Material	Category	Cr	Ni	Mo	N	C	Si	P	S	Cu	Others
6 % Mo Alloy 1	Austentic	20.8	17.4	6.0	0.206	0.012	0.42	0.025	<0.003	0.69	Co 0.24 Nb 0.04
6 % Mo * Alloy 2	Austentic	20.7	24.8	6.24	0.134	0.014	0.29	0.022	0.003	0.80	–
25 % Cl, 5 % N, stainless steel	Duplex	24.4	6.7	4.0	0.22	0.036	0.31	0.019	<0.003	0.55	W 0.80

*The nitrogen content of this material has recently been increased by the manufacturer; material to this analysis is no longer produced.

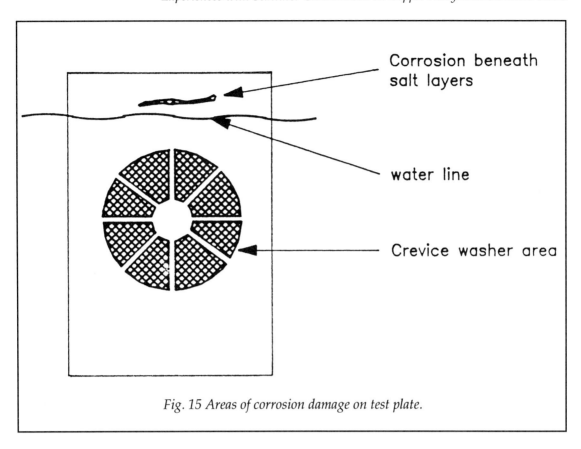

Fig. 15 Areas of corrosion damage on test plate.

accepted theory that pit initiation becomes easier at higher temperatures. There is no evidence of pit depth increasing with temperature, in fact, for two of the alloys tested there is an apparent decrease. This lower pit propagation rate at higher seawater temperatures has been seen by other workers [5] for the alloys AISI 304 and AISI 316 stainless steel. The effect is considered to be due to a combination of:

(a) the decrease in oxygen solubility with increased seawater temperature;
(b) an increase in general corrosion rate of the bare, uncreviced surface (a competing anodic reaction);
(c) alterations in the nature of the surface film/biofilm.

The monitoring of potential of the samples during the 100 day exposure period indicated that for all the samples a wide potential range was possible (–600 to +700 mV SCE). An example of the potential variation is shown in Fig. 18, where the potentials for the duplex alloy are given for the three test temperatures at a chlorination level of 1.3 ppm. Strong similarities between the variation of potentials of samples and minor temperature or chlorination level changes confirm previous observations that the potential is far more sensitive to small changes in bath temperature and chlorination than to the degree of localised corrosion. Thus, monitoring of potential of these alloys does not appear to be a very useful tool in predicting corrosion damage.

The corrosion damage beneath salt layers was only temperature dependent; there was no correlation with chlorination level. This is to be expected, since the moist salt layers are mainly formed as a result of evaporation and condensation whereupon chlorine present in the seawater would be lost.

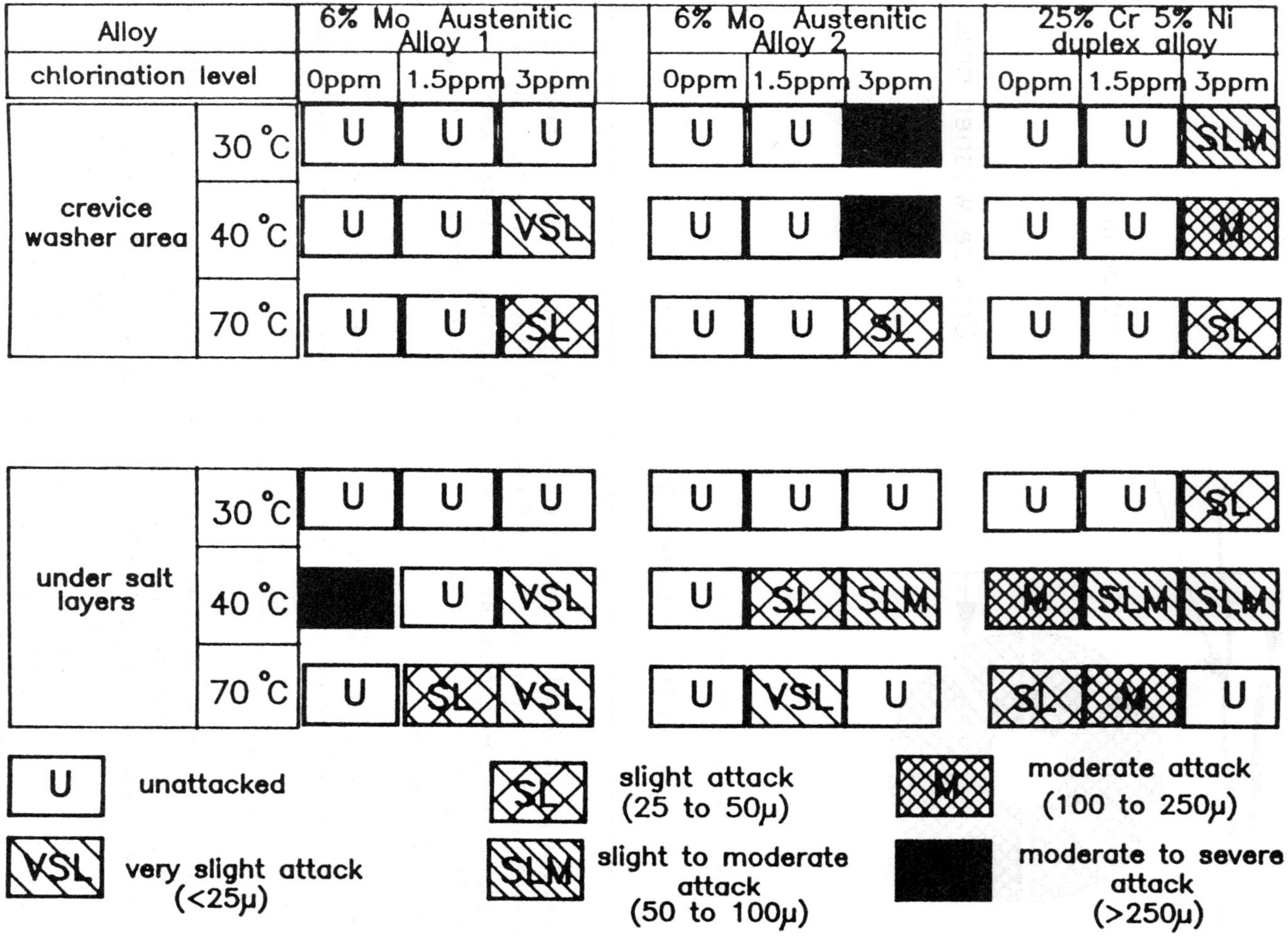

Fig. 16 Summary chart depicting the extent of corrosion attack for all the combinations of temperature, alloy and chlorination levels tested.

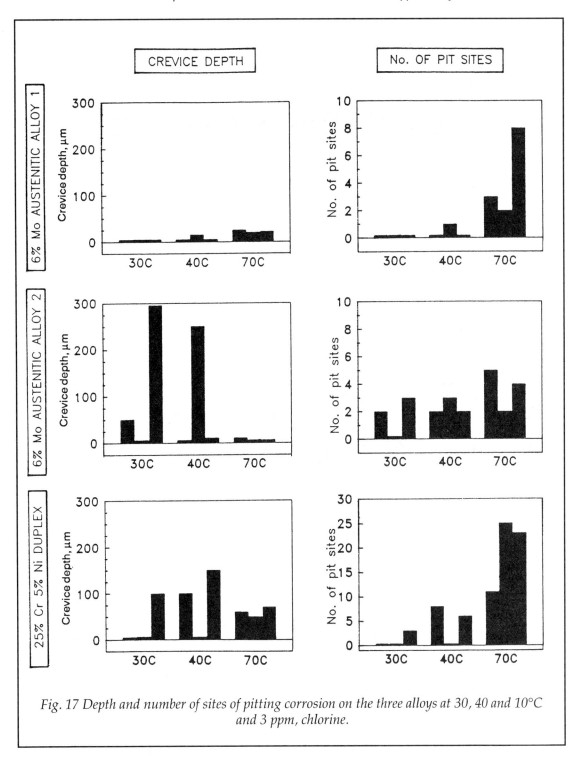

Fig. 17 Depth and number of sites of pitting corrosion on the three alloys at 30, 40 and 10°C and 3 ppm, chlorine.

7. Conclusion

It is apparent that when studying the behaviour of materials for use in seawater systems, the effect of chlorine, even at low levels, can be extremely significant, both in its effect as a biocide in minimising the formation of the 'catalytic' biofilm layer and in a role as an oxidising agent.

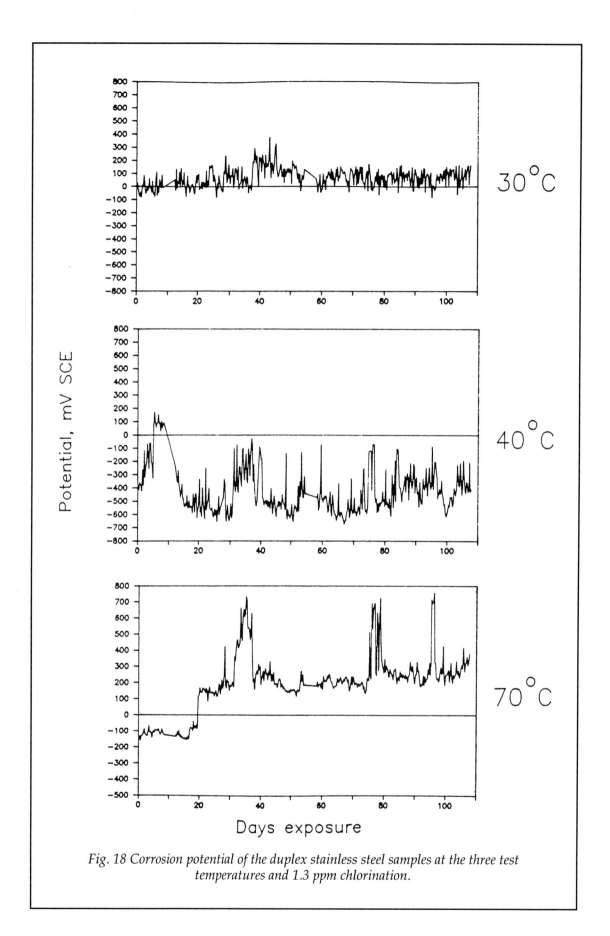

Fig. 18 Corrosion potential of the duplex stainless steel samples at the three test temperatures and 1.3 ppm chlorination.

References

1. E. B. Shone and G. C. Grim, 25 years experience with seawater cooled heat transfer equipment in the Shell Fleets, Trans. I. Mar. Eng., **98**, paper 11.
2. P. Gallagher, R. E. Malpas and E. B. Shone, Corrosion of stainless steels in natural, transported and artificial seawaters, Br. Corros. J., 1988, **23**, (4), 229–233.
3. V. Scotto, R. Di Cinitio and G. Marcenero, The influence of marine aerobic microbial films on stainless steel corrosion behaviour, Corros. Sci., 1985, **25**, 185.
4. B. Wallen, Effect of chlorination on stainless steels in seawater. Corrosion NACE '88, paper 403.
5. R. M. Kain and T. S. Lee, Crevice corrosion of stainless steels in ambient and elevated temperature seawaters, 5th Int. Congr. on Marine Corrosion and Fouling, Barcelona, 1980.

7

Corrosion of Stainless Steels Caused by Bromine Emission from Chlorinated Seawater

J.W. OLDFIELD AND B. TODD*

Cortest Laboratories Ltd, UK
*Consultant to Nickel Development Institute

1. Introduction

The desalination industry has developed rapidly in the last 20 years and is now a major user of corrosion-resistant materials. About 70% of existing capacity uses the multi-stage flash (MSF) process. This is a heat exchange process in which seawater is made to boil(flash) in many stages at progressively lower temperatures and pressures. The pressures in the plant are controlled by an ejector system. This system is usually made from Type 316 stainless steel— chosen for its resistance to the normal mixture of gases evolved from seawater i.e. oxygen, carbon dioxide and water vapour. However, in some plants serious corrosion has occurred in this vent system. These problems have been caused by bromine being evolved from chlorinated seawater. The purpose of this paper is to define the conditions under which bromine emission can occur and to describe the types of corrosion which this can cause.

2. Water Treatment in MSF Plants

Chlorination is normally used to control marine fouling in these plants. This is controlled by the residual at the outlet—a value of 0.1–0.2ppm is often used.

To control scaling from the calcium and magnesium salts in seawater two methods are used:

(i) Acidification of the feed to decompose carbonates and bicarbonates.

(ii) Addition of polyphosphates or organic polymers which inhibit scaling.

When acid dosing is used, the pH of the seawater feed is reduced to a pH of between 4–5. It is then passed through a decarbonator to remove the carbon dioxide which raises the pH to 5–6. It is then deaerated. The early failures in stainless steel vent systems were all in acid dosed plants.

With additive dosing there is no fall in pH. This remains at about the normal seawater level of pH8.

3. Chemistry of Chlorine in Seawater

The chemistry of the reactions of chlorine in seawater have been studied in Ref. [1]. This shows that the 70 ppm bromide normally present in seawater reacts with residual chlorine. This reaction goes virtually to completion in reducing the chlorine and oxidising the bromides to bromine. At normal seawater pH most of the residual halogen is present as hypobromous acid and hypobromite ion(HBr + HBrO).

However, in acid-dosed plants where the pH is reduced to as low as 4, the bromine species

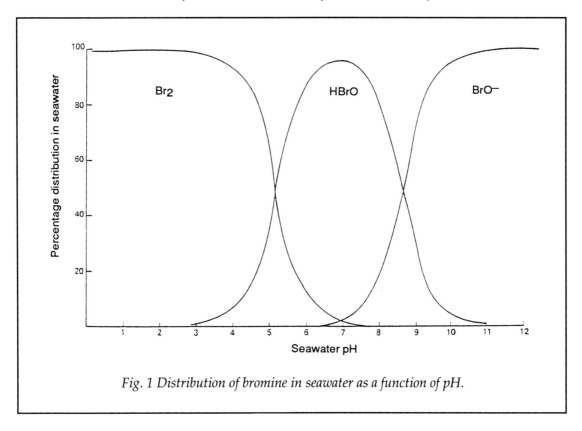

Fig. 1 Distribution of bromine in seawater as a function of pH.

present will contain bromine gas in solution. Figure 1 shows the effect of these changes in pH on the bromine compounds present from which it can be seen that, at pH4, most of the residual halogen is present as bromine gas in solution. Even when the pH rises during decarbonation to pH5–6 there will still be a significant proportion of the residual present as bromine in solution.

4. Vent System Corrosion in Acid-dosed Plants

When low pH seawater is deaerated the dissolved bromine is stripped out with the other gases and passes into the stainless steel venting system. Also, the mass flow of seawater through the deaerator is 1000–10 000 times that of the gases. The bromine is concentrated by this factor. Thus, the low levels in the seawater can become high levels in the vent system and any condensates from these gases will have a much higher bromine content than the seawater. Figure 2 shows the effect of seawater pH on the bromine concentration in the vent gases at three chlorine levels. Figure 3 shows the bromine levels in condensates from the vent gases.

Corrosion of stainless steel in acid-dosed plant vent systems is characterised by a multiplicity of fine pits which eventually lead to perforation. Cracking is normally absent and analyses of corrosion products show chlorides to be virtually absent.

Work on the effect of bromine on corrosion of stainless steel shows that attack is likely above 10ppm[2].

Figure 3 shows how, by controlling residual chlorine and pH at the deaerator inlet, the bromine level can be controlled to an acceptable level. If for process reasons it is not acceptable to obtain the levels indicated in Fig. 3 then the feed can be dechlorinated with sodium sulphite prior to deaeration [3].

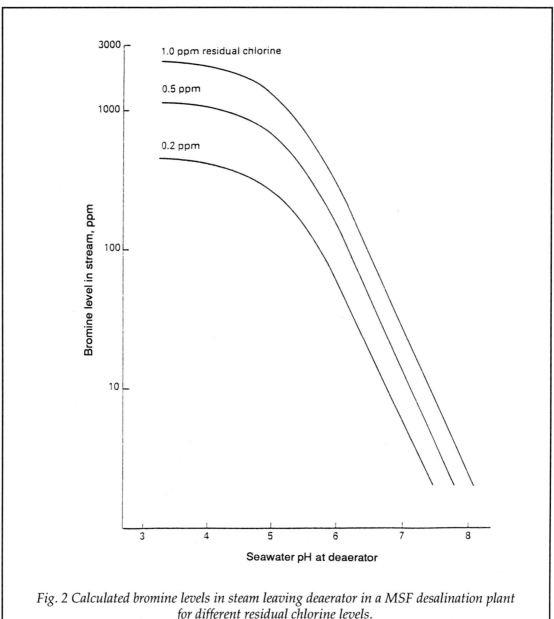

Fig. 2 Calculated bromine levels in steam leaving deaerator in a MSF desalination plant for different residual chlorine levels.

5. Vent System Corrosion in Additive Plants

As described earlier, these plants operate at about pH 8 and so should not evolve bromine by the process described above. However, corrosion problems caused by bromine have been experienced in several MSF units in the Middle East [4]. In these cases the corrosion first manifested itself as external cracking but when samples were removed for examination the cracked area had suffered severe general corrosion. An analysis of corrosion products gave the following results: bromin —3.3%, chlorine—0.14%.

The presence of bromine (as bromides at this stage) was unexpected but the different characteristics of the corrosion and the significant chlorine level indicated a mechanism of evolution different from that described above.

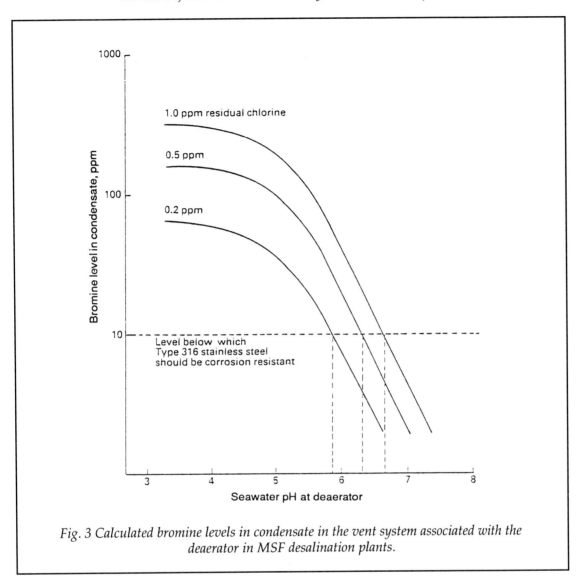

Fig. 3 Calculated bromine levels in condensate in the vent system associated with the deaerator in MSF desalination plants.

In one of the affected units it was noted that the seawater contained above-normal levels of ammonia. An analysis from a vent condenser drain gave the following data: free chlorine (probably bromine)—0.2ppm; ammonia—0.3ppm.

Metallographic examination of the cracks in the stainless steel showed that they were typical transgranular multi-branched stress corrosion cracks (SCC) as normally seen with chloride SCC.

It is known that bromine and chlorine can react with ammonia to form bromamines and chloramines as follows:

$$NH_3 + HOBr = NH_2Br + H_2O$$

$$NH_2Br + HOBr = NHBr_2 + H_2O$$

$$NHBr_2 + HOBr = NBr_3 + H_2O$$

and that similar reactions occur with chlorine.

Experimental data indicate that the reaction time to form chloramines is of the order of minutes and would not be complete before the chlorine was reduced by the bromide ions present. Hence in chlorinated seawater bromamines would predominate, which agrees with the corrosion product given earlier.

The actual bromamine formed depends on the ratio of bromine to ammonia in the seawater although significant amounts of chloramine may form. Reference [5] indicates that at a 1:1 ammonia:chlorine ratio in terms of ppm (5:1 in terms of molarity) about 10% of the total oxidant present is monochloramine at pH 8 and 25°C. Once formed, bromamines (and chloramines) can decompose as follows:

$$4NH_2Br + H_2O = N_2 + 3HBr + 2NH_3$$

$$2NHBr_2 + H_2O = N_2 + 3HBr + HOBr$$

$$2NBr_3 + H_2O = N_2 + 3HBr + 3HOBr$$

The cause of the severe general corrosion noted on the stainless steel in the venting system is probably the formation of hydrobromic and hypobromous acids

In order to investigate the possibility of the above mechanisms for stripping bromine from seawater, laboratory experiments were performed with a 3.5% sodium chloride solution containing bromine and ammonia. This was degassed using argon as a stripping gas. The evolved gases were absorbed in a pH 10.5 solution of caustic soda at ambient temperature. The tests were carried out at 40 and 90°C to simulate conditions at the high and low temperature sections of an MSF plant. The gases absorbed by the caustic soda were analysed using ultra-violet absorption.

Figure 4 shows the results of these experiments which demonstrate that bromamines can be evolved from simulated seawater at both 40 and 90°C. Also, evolution continued for over 1 hour—a long time compared with that taken for seawater to flow through the plant. It is possible therefore that bromamines were evolved at any (or all) stage of the process. The fact that corrosion is confined to the high temperature section of the venting system may be because bromamine decomposition occurs more readily there or because the concentration of bromine compounds increases as it passes through the venting system.

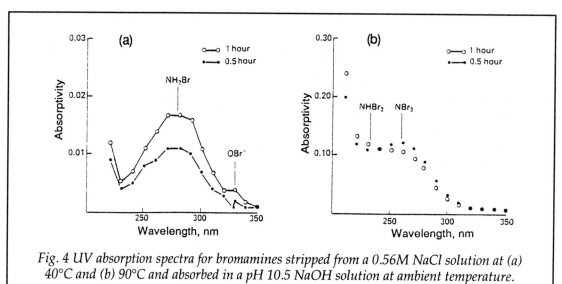

Fig. 4 UV absorption spectra for bromamines stripped from a 0.56M NaCl solution at (a) 40°C and (b) 90°C and absorbed in a pH 10.5 NaOH solution at ambient temperature.

6. Summary and Conclusions

When chlorine is added to seawater bromine compounds are normally formed. In acid dosed MSF plants the reduction in pH converts much of the residual chlorine to bromine gas. This gas can be stripped in the deaerator, giving pitting corrosion in the stainless steel venting system.

For additive plants, the presence of ammonia in the seawater can lead to the formation of bromamines and chloramines which can be stripped from the seawater to concentrate and decompose in the venting system. The cracking found in this case may be due to the chlorine carried over by the chloramines or it may be because one of the bromamines is a stress corrosion cracking agent. The severe general corrosion is probably caused by the acids formed by the bromamine and chloramine decomposition.

These failures emphasise the need to minimise chlorine residuals, maintaining just sufficient to control marine growth.

References

1. J. W. Oldfield and B. Todd, Desalination, 1981, **38**, 233.
2. P. E. Morris and W. H. Wendeler, INCO Internal Report.
3. N. Nada, NWSAIA Conference, Washington, 19481.
4. W. S. S. Lee, J. W. Oldfield and B. Todd, 1st World Congr. on Desalination and Water Re-Use, Florence, May 1983, pp. 209–221.
5. R. Sugam and G. R. Helz, Chemosphere, 1981, **10**, 41.

Testing

8

Improved Method for Measuring Polarisation Curves of Alloys during Prolonged Times of Exposure

F. P. IJsseling

Corrosion laboratory, Royal Netherlands Naval College (Harssens), c/o Marinepostkamer,
PB 10.000, 1780 CA Den Helder, Netherlands

Abstract

An electrochemical cell is described, which can be attached to a large plate sample, allowing measurements to be performed on a small, well-defined part of the surface, without affecting the plate surface outside the cell. This method has several advantages, e.g. the occurrence of artefacts due to previous measurements can be avoided. The method is demonstrated for a stainless steel exposed during a prolonged time in flowing seawater, showing the change in polarisation behaviour due to the formation of a biofilm.

1. Introduction

In many cases it is desirable to measure possible changes in the polarisation behaviour of alloys exposed for long periods of time. An example of this are stainless steels, which exhibit changing polarisation characteristics due to the formation of a microbial slime layer. This can be observed by a dramatic increase of the free corrosion potential (Fig. 1), which is linked directly to changes in the partial anodic and/or cathodic current densities. The main action of the biofilm is to enhance the cathodic reaction. This problem has been addressed in several papers; polarisation curves have been used to explain the underlying causes [1–9].

However, it was felt that the experimental method of measuring polarisation curves to determine the effect of time due to surface film formation could be improved by the development of a special type of electrochemical cell.

2. Experimental

For long term exposure, during which polarisation curves have to be determined, either large plate samples can be used or a number of small electrodes, which are discarded after use. The latter may show severe edge effects, which in the case of stainless steels may seriously affect the results of the polarisation measurements and so for this application large plate samples are to be preferred. However, large samples also may have some serious drawbacks:

(i) it is difficult to obtain a homogeneous current distribution over the surface during polarisation, and

(ii) in particular, when it is intended to perform a number of consecutive measurements on the same sample there is the danger of artefacts due to irreversible changes of the surface and the surface layer caused by the previous electrochemical polarisation.

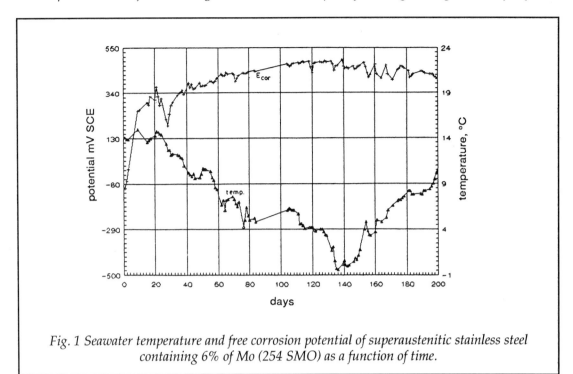

Fig. 1 Seawater temperature and free corrosion potential of superaustenitic stainless steel containing 6% of Mo (254 SMO) as a function of time.

To prevent such undesirable side effects a special cell was constructed. Generally the following requirements can be given for an electrochemical cell:

(i) homogeneous current distribution over the surface of the working electrode;

(ii) potential measurement with minimum error due to iR-drop between the working electrode surface and reference electrode;

(iii) constant mass transfer conditions at the working electrode, and

(iv) no errors due to accumulation of reaction products at the electrodes.

The cell was developed from a probe used for *in situ* polarisation tests, e.g. cathodically protected submerged structures [10].

The cell can be attached to the surface of a large plate sample by means of vacuum (Fig. 2). The circular surface selected for polarisation has a diameter of 2 cm. The plate sample was placed horizontally in a glass vessel with a continuous supply of fresh seawater (Fig. 3(a)). The seawater flow was directed to the outer PVC tank only during actual polarisation, at the same time the water level in the glass vessel was lowered to prevent electrolytic contact between the seawater in both tanks (Fig. 3(b)). Using this arrangement, two possible problem areas were avoided:

(i) electrolytic contact between the working electrode inside the cell and the remaining plate surface outside;

(ii) earthing problems with the potentiostatic equipment (N.B: a potentiostat is now commercially available which is fully floating, thus simplifying electrochemical measurements in flowing solutions connected to earth).

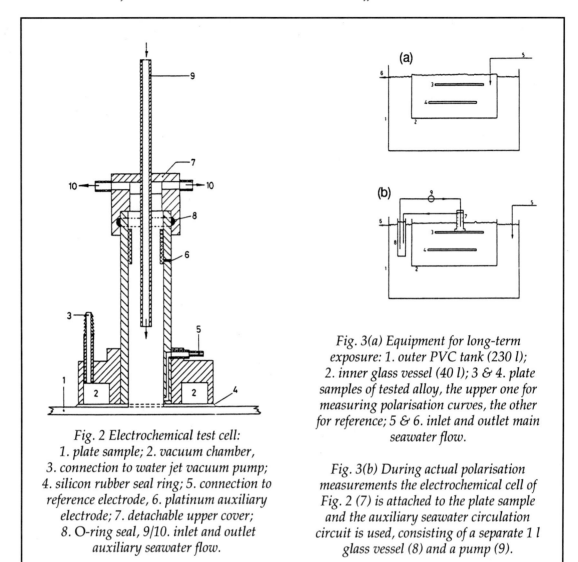

Fig. 2 Electrochemical test cell:
1. plate sample; 2. vacuum chamber, 3. connection to water jet vacuum pump; 4. silicon rubber seal ring; 5. connection to reference electrode, 6. platinum auxiliary electrode; 7. detachable upper cover; 8. O-ring seal, 9/10. inlet and outlet auxiliary seawater flow.

Fig. 3(a) Equipment for long-term exposure: 1. outer PVC tank (230 l); 2. inner glass vessel (40 l); 3 & 4. plate samples of tested alloy, the upper one for measuring polarisation curves, the other for reference; 5 & 6. inlet and outlet main seawater flow.

Fig. 3(b) During actual polarisation measurements the electrochemical cell of Fig. 2 (7) is attached to the plate sample and the auxiliary seawater circulation circuit is used, consisting of a separate 1 l glass vessel (8) and a pump (9).

During polarisation measurements the outer tank acted as a temperature buffer for the inner tank. The cell was provided with an internal platinum auxiliary electrode, which was positioned symmetrically with respect to the working electrode to promote a homogeneous current distribution over the sampled surface. A small bore channel which ended closely to the plate surface was used to connect the reference electrode, thus limiting the uncompensated resistance between the tip of the reference and the working electrode.

Inside the cell a separate seawater flow was directed over the sampled electrode surface and the auxiliary electrode in order to:

- sweep away possible reaction products resulting from the electrochemical polarisation; and
- maintain constant mass transfer conditions at the surface of the working electrode area.

This seawater was contained in a separate auxiliary circuit which was also electrically isolated from the main seawater supply. By measuring polarisation curves at a new location of the cell on the plate surface every time, the effect of the history of the sampled surface was negligible and artefacts due to previous measurements were avoided. The polarisation curves were measured potentiodynamically, starting from the free corrosion potential. The anodic and cathodic curves were measured separately, using a fresh location for each curve.

All experiments were performed using stainless steel 254 SMO, exposed to flowing natural seawater in a tank as shown in Fig. 3.

3. Results and Discussion

During preliminary experiments the cell was polarized from −500 to +250 mV while simultaneously measuring the potential of the plate sample outside the cell against a second reference electrode (Fig. 4). The potential at the outside changed only a little during polarisation, and these changes were possibly the result of capacitive influences.

The influence of artefacts was demonstrated by measuring the anodic curve after previous cathodic polarisation at the same location, notwithstanding a pause between the measurements of 20 h (Fig. 5(a)). A similar effect was found on polarizing in the cathodic direction if an anodic curve had been measured previously (Fig. 5(b)). Moreover, scan speed was found to have a significant effect on the shape of the polarisation curves (Fig. 6), applying scan rates of 0.9–0.009 mVs^{-1}. All subsequent curves were measured with a constant scan rate of 0.09 mVs^{-1}.

The reproducibility of the measurements is quite good: in Fig. 7 a number of polarisation curves are shown which were obtained after cleaning, freshly abrading and exposing for 16h, respectively, the plate sample. After formation of a thicker biofilm the reproducibility decreased mostly because of the more or less undefined character of the maxima in the cathodic part of the polarisation curve (see Fig. 6(a)).

Finally, in Fig. 8 the polarisation curves are shown as measured after increasing times of exposure. It can be seen that the cathodic part of the curves is affected in particular, the cathodic reaction proceeding faster and at more noble potentials. So the main changes in the overall polarisation behaviour are found between −100 and +300 mV (SCE), as is also indicated by the increase of the free corrosion potential upon exposure to natural seawater. This is in agreement

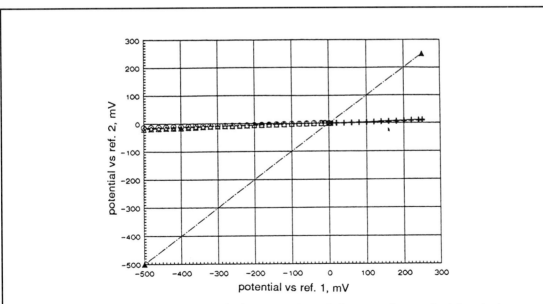

Fig. 4 Shift of free corrosion potential of outer surface of plate sample on polarising the inner surface from −500 to +250 mV vs SCE.

with the measurements and the suppositions made in the references mentioned before [1–9].

The development of the cathodic current values with time are also in accord with the galvanic current and couple potential of 254 SMO galvanically coupled to CuNi10Fe (Fig. 9)

As such these measurements do not give a complete insight into the underlying cause of the change of the cathodic polarisation behaviour on biofilm formation. To obtain this goal the application of more extensive methods, including biochemical, analytical and surface–analytical techniques, must be considered.

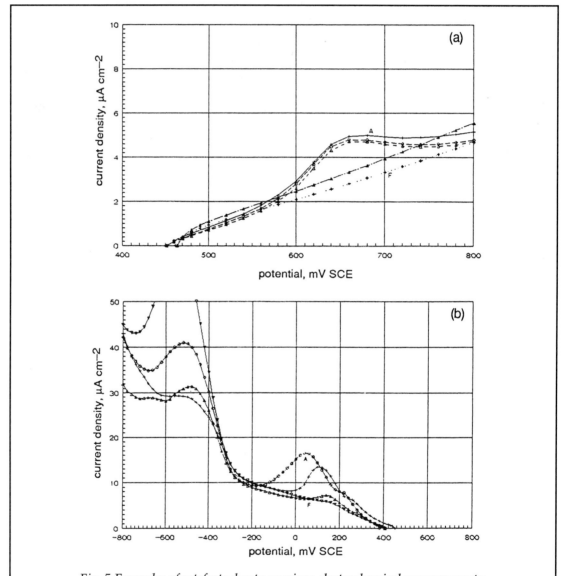

Fig. 5 Examples of artefacts due to previous electrochemical measurements: (a) anodic polarisation curves of 254 SMO after 140–150 days of exposure, showing artefacts:

(b) similarly measured cathodic polarisation curves, showing artefacts after potentiodynamic polarisation to +800 mV (scan speed 0.09 mVs^{-1}) which had been performed 20 h before. A - measured at location which 20 h before had been polarised potentiodynamically with a scan speed of 0.09 mVs^{-1} in the cathodic direction to −800 mV; F - measured at fresh, hitherto only freely exposed location.

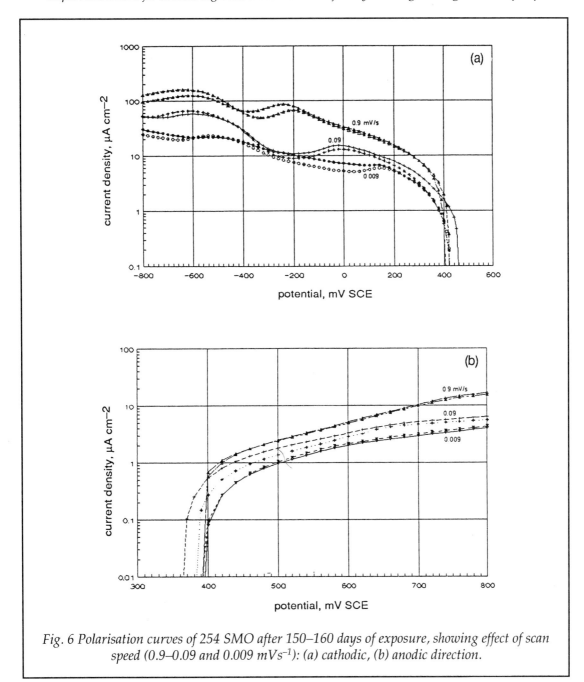

Fig. 6 Polarisation curves of 254 SMO after 150–160 days of exposure, showing effect of scan speed (0.9–0.09 and 0.009 mVs^{-1}): (a) cathodic, (b) anodic direction.

However, it has been shown that with this type of cell it is certainly possible to carry out electrochemical measurements on large plate samples.

4. Conclusions

An electrochemical cell has been developed, to be used in combination with large plate samples to undertake electrochemical polarization measurements locally at well-defined spots, without damaging or influencing the surrounding plate surface. In this way the occurrence of artefacts due to previous electrochemical polarization is prevented. In a series of experiments performed on stainless steel samples the cell came up to expectations, providing a better

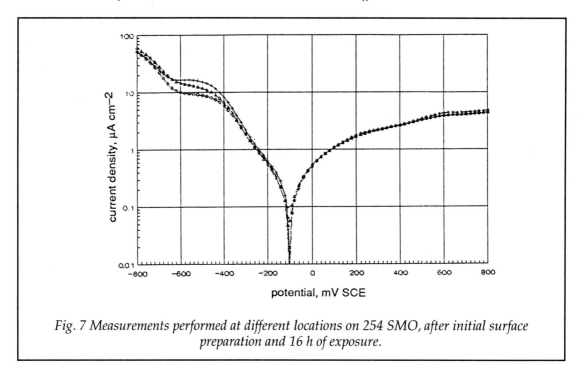

Fig. 7 Measurements performed at different locations on 254 SMO, after initial surface preparation and 16 h of exposure.

insight into the development of the polarization characteristics on exposure to seawater during prolonged periods of time.

Acknowledgements

The support of Avesta AB, Sweden, in providing the samples, is gratefully acknowledged.

References

1. A. Mollica, A. Trevis, E. Traverso, G. Ventura, V. Scotto, G. Alabiso, G. Marcenaro, U. Montini, G. de Carolis and R. Dellepiane, Proc. 6th Int. Congr. on Marine Corrosion and Fouling, Athens, 1984, 269.
2. V. Scotto, R. Di Cinto and G. Marcenaro, Corros. Sci., 1985, **25**, 185.
3. R. Johnsen and E. Bardal, NACE Corrosion'86, Houston, Tx, paper 227.
4. V. Scotto, G. Alabiso and G. Marcenaro, Bio-electrochemistry and Bio-energetics 1986, **16**, 347 (section of J. Electroanal. Chem., constituting Vol. 212, 1986).
5. R. Holthe, P. O. Gartland and E. Bardal, Proc. 7th Int. Congr. on Marine Corrosion and Fouling, Valencia, 1988.
6. S. C. Dexter and G. Y. Gao, Corrosion, 1988, **44**, 717.
7. R. Holthe, The Cathodic and Anodic Properties of Stainless Steels in Sea Water, Thesis, Univ. of Trondheim, 1988.
8. O. Varjonen, T. Hakkarainen, E.-L. Nurmiaho-Lassila and M. Salkinoja Salonen, Microbial Corrosion—1, Proc. 1st European Workshop on Microbial Corrosion, Portugal, 1988, p. 164, Elsevier Science Publ. Ltd, London, 1988.
9. A. Mollica, A. Trevis, E. Traverso, G. Ventura, G. de Carolis and R.Dellepiane, Corrosion, 1989, **45**, 48.
10. R. D. Strommen, H. Osvoll and W. Keim, NACE Corrosion '86, Houston, Tx, paper 297.

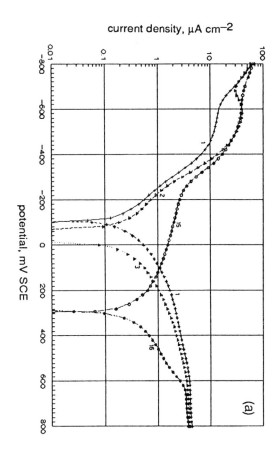

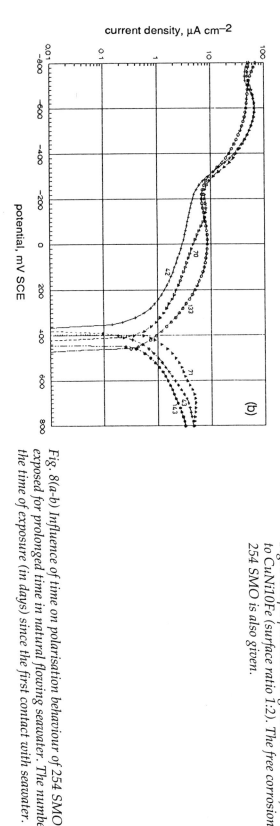

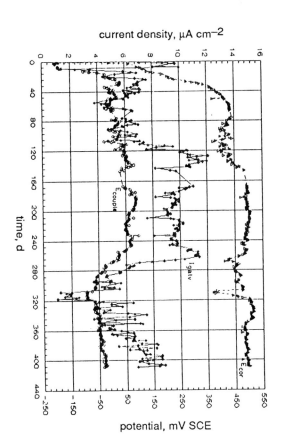

Fig. 8(a-b) Influence of time on polarisation behaviour of 254 SMO, when exposed for prolonged time in natural flowing seawater. The numbers indicate the time of exposure (in days) since the first contact with seawater.

Fig. 9 Couple potential and galvanic current of 254 SMO coupled to CuNi10Fe (surface ratio 1:2). The free corrosion potential of 254 SMO is also given.

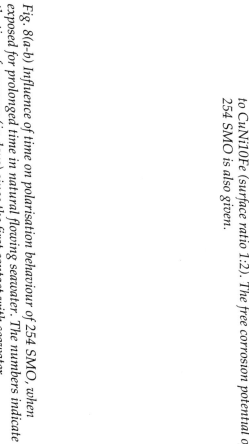

Application of Electrochemical Impedance Spectroscopy to Monitor Seawater Fouling on Stainless Steels and Copper Alloys

D. Féron

Commissariat à l'Energie Atomique, CEREM/SCECF, Etablissement COGEMA de la Hague,
50444 Beaumont-Hague, France

Abstract

Electrochemical impedance spectroscopy may be applied to detect and to follow seawater fouling. Experiments have been conducted with natural seawater flowing inside tube electrodes at temperatures between 30 and 85°C. With stainless steel tubes, the development of mineral and organic foulings has been monitored; a linear relationship between the dry weight of the organic fouling and its electrical resistance has been observed. On copper alloy tubes, only mineral deposits have occurred and have been detected by impedance spectroscopy.

1. Introduction

In components exposed to seawater, the mineral and organic deposits on surfaces are known as 'fouling'. These deposits decrease heat transfer, increase fluid frictional resistance, affect the efficiency of cathodic protection and accelerate corrosion processes. For instance, the presence of microbiological film (called 'slime' or 'biofilm') is often put forward to explain that corrosion in natural environments is higher than in artificial and sterile conditions.

The objective of this paper is to show that Electrochemical Impedance Spectroscopy (EIS) may be applied to detect and evaluate seawater fouling (organic and mineral deposits).

2. Background

EIS is a technique which has many applications in corrosion research, and which is now commonly carried out in order to collect information relating to the metal/solution interface.

EIS may be successfully applied to seawater fouling if this can be characterised by (i) an electrical resistance, R_f and (ii) an electrical capacitance, C_f.

The addition of these fouling parameters to the parameters of the metal/solution interface leads to the equivalent circuit shown in Fig. 1. With such a model, the Nyquist complex plane plot is made up of two semi-circles, as shown on Fig. 1, with time constants given by:

$$t_f = R_f \cdot C_f \text{ and } t_m = R_t \, C_d$$

The semi-circle occurring at higher frequencies will be caused by seawater fouling if t_f is higher than t_m. The shape of the plot will appear as two distinct semi-circles if the following criteria are met:

$$0.2 < R_f/R_t < 5 \text{ and } t_m/t_f > 20$$

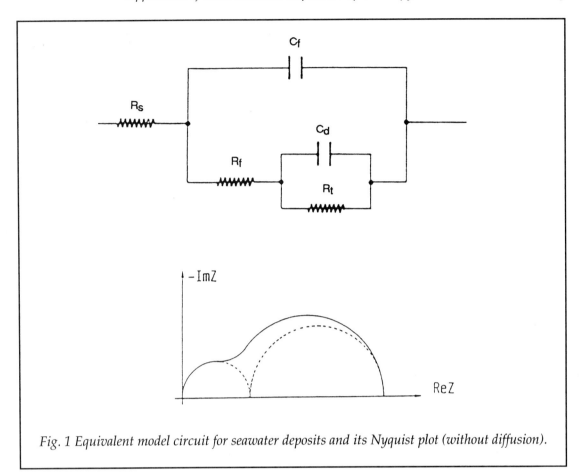

Fig. 1 *Equivalent model circuit for seawater deposits and its Nyquist plot (without diffusion).*

These values are probably questionable, but they clearly indicate that the evidence of fouling (R_f) will be correlated with the corrosion behaviour (R_t). For instance, with titanium (which is virtually immune to corrosion in seawater), R_t values are very high and so the first criterion will be met only when R_f values are also high, and only when deposits are thick enough.

3. Experimental

Experiments have been conducted in the SIRIUS facility which is located at La Hague (Normandy/France). SIRIUS is a single pass loop which operates with natural seawater at temperatures between 20 and 150°C.

Electrochemical measurements have been made on tube-electrodes (internal diameter: 17 mm, length: 50 mm). Seawater flows inside these tube-electrodes at 2 ms^{-1}. An electrolytic bridge connects the seawater flowing inside to an external saturated calomel reference electrode (Fig. 2).

Impedance measurements have been performed at the free corrosion potential using a potentiostat and a frequency response analyser (Solartron Schlumberger 1286 and 1250).

4. Results and Comments

The experiments have included titanium, stainless steel and copper alloy electrodes covered by various seawater deposits.

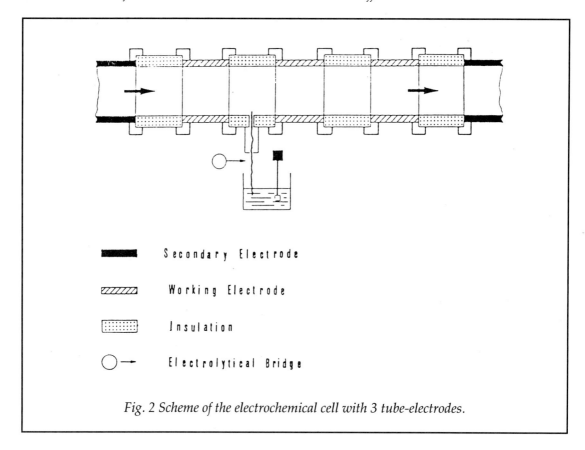

Fig. 2 Scheme of the electrochemical cell with 3 tube-electrodes.

In order to illustrate the results obtained with copper alloys, two EIS diagrams are shown on Fig. 3. These were obtained at 80°C, after 20 h of exposure. A 3% NaCl solution flowed inside the tube-electrode for the first diagram (Fig. 3(a)), and the observed relaxations were due only to corrosion phenomena. With natural seawater, only one relaxation has been observed around 1 Hz, and this increased with time; it was due to the formation of mineral deposits which occur under these conditions (mainly hydroxides above 70°C). The fouling resistance calculated from the EIS diagram in Fig. 3(b) (R_f= 850 Ω/cm^{-2}) correlated with the weight of the mineral deposits (0.2 mg cm^{-2}). By following the values of the fouling resistance, the increase in the precipitation of hydroxides (mainly $Mg(OH)_2$) has been pointed out up to 85°C as shown in Fig. 4.

On copper alloy tube-electrodes, no fouling (neither organic nor mineral) occurred at the low temperature (35°C).

On stainless steels, organic fouling has been detected and followed by EIS. Figure 5 illustrates the results obtained with fresh and natural seawater flowing inside the tube-electrodes at 2 ms^{-1}. At the beginning of the exposure or when the dry weight of the deposits was lower than 1 mg cm^{-2}, no relaxation was visible on the Nyquist diagram; when the fouling increased, a relaxation appeared around 100 Hz and increased with time as shown on Fig. 5(b) and 5(c). At the end of the exposure time, the electrical fouling resistance was measured by EIS and plotted vs the dry weight of the deposits found on the electrodes: as shown on Fig. 6, a linear relationship was observed.

Results on titanium were similar to those obtained on stainless steel except that detection of smaller deposits was difficult. For instance, after 3000 h of exposure at 35°C, the EIS diagram was characteristic of a passive metal with low corrosion rate (as shown in Fig. 5(a)): it shows high R_t, but no clear evidence of fouling when the dry weight of organic deposits is 1.9 mg cm^{-2}.

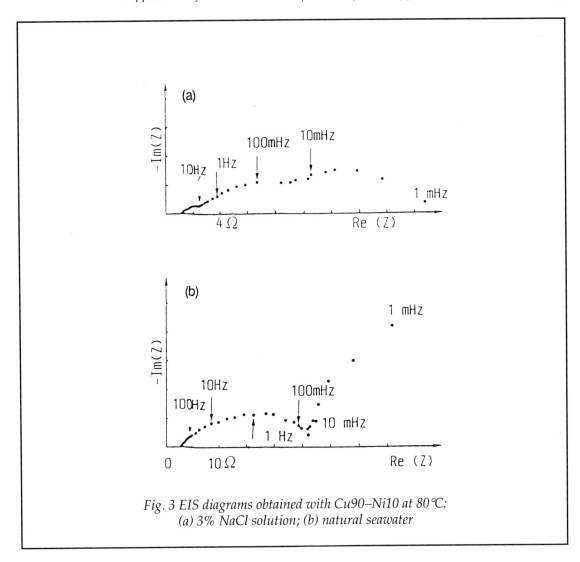

Fig. 3 EIS diagrams obtained with Cu90–Ni10 at 80°C:
(a) 3% NaCl solution; (b) natural seawater

This interpretation of EIS data using a linear coordinate plot in the complex plane (Nyquist plot) is a first step. Of course, more information and lower detection limits are available using the Bode plot and mathematical simulation. The purpose of this presentation was only to show that EIS may be useful for the detection and the evaluation of seawater fouling.

5. Conclusion

The detection and the evaluation of seawater fouling has been performed on tube electrodes at temperatures between 35 and 85°C.

The fouling evidence by EIS is not only a function of the nature and of the thickness deposits, but also of the materials on which fouling occurs.

From these experiments, Electrochemical Impedance Spectroscopy appears in be helpful in detecting and following seawater fouling.

References

1. N. J. Dowling, M. W. Mittleman and J. C. Danko, Microbially Influenced Corrosion and Biodeterioration, The University of Tennessee, Knoxville, Tennessee, USA, 1991.

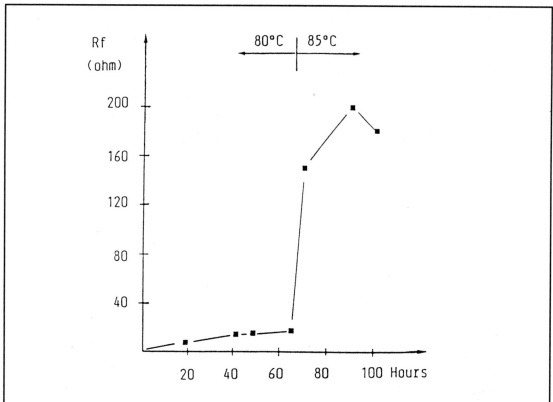

Fig. 4 Development of the electrical fouling resistance (R_f) on copper–nickel tube during temperature increase (natural flowing seawater, 2 ms^{-1}).

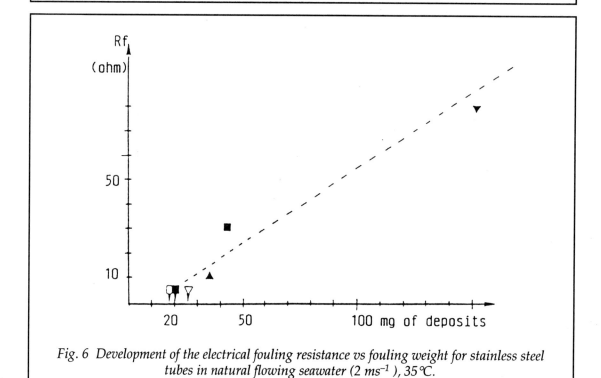

Fig. 6 Development of the electrical fouling resistance vs fouling weight for stainless steel tubes in natural flowing seawater (2 ms^{-1}), 35°C.

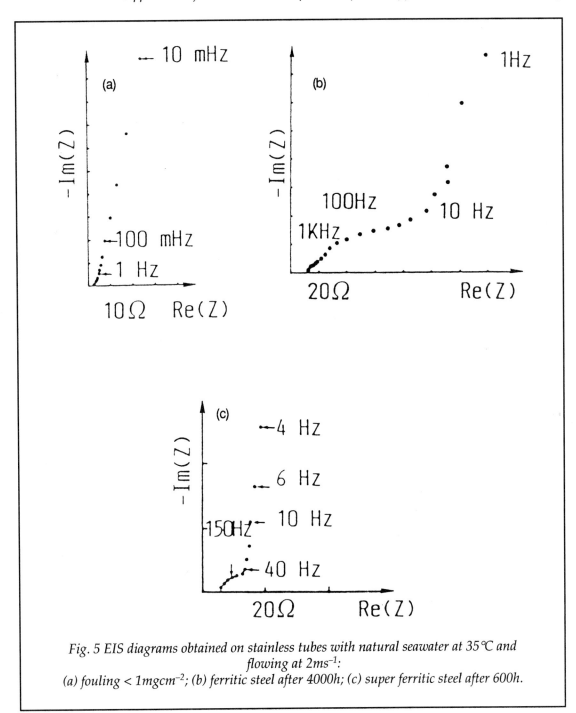

Fig. 5 EIS diagrams obtained on stainless tubes with natural seawater at 35°C and flowing at 2ms⁻¹:
(a) fouling < 1mgcm⁻²; (b) ferritic steel after 4000h; (c) super ferritic steel after 600h.

2. Spécial Biocorrosion, Materiaux et Techniques, Paris, December 1990.
3. Proc. 1st Int. Symp. on Electrochemical Impedance Spectroscopy, Bombannes, France, 22–26 May, 1989.
4. Deuxième Forum sur les Impédances Electrochimiques, Montrouge, France, 28–29 October 1987.
5. Proc. 1st European Workshop on Impedance Measurements, Karlsruhe, Germany, 8–9 April, 1987.

Effect of Temperature on Initiation, Repassivation and Propagation of Crevice Corrosion of High-Alloy Stainless Steels in Natural Seawater

S. Valen, P. O. Gartland and U. Steinsmo

Sintef Corrosion Center, N-7034 Trondheim, Norway

Abstract

Experiments have been performed to study initiation and repassivation temperatures of the high-alloy austenitic stainless steel Avesta 254 SMO and the high-alloy duplex stainless steel Sandvik SAF 2507. The results show that high-alloy stainless steels have a critical initiation temperature, T_{init}, and a lower temperature for repassivation, T_{repass}. Effects of potential and crevice-forming material on T_{repass} are discussed. To evaluate the propagation of crevice corrosion, results from long time anodic and cathodic polarisation of separate anode and cathode specimens are used as boundary conditions in a computer programme, GALVCORR [1], to calculate potential and corrosion rate profiles in piping systems made of Avesta 254 SMO. Effects of pipe geometry and temperature on the crevice corrosion rate are discussed.

1. Introduction

In recent years, high-alloy stainless steels have been commonly used in seawater systems. Crevice corrosion is responsible for most of the failures. For temperatures below 30°C, these new steels are very resistant against initiation of crevice corrosion. Although 30°C is specified as an upper limit, crevice corrosion is observed on high-alloy stainless steels at lower temperatures [2]. Crevice corrosion may either initiate in weak spots in welds or cast material in flange connection, or the service temperature may temporarily be higher than specified. When the temperature is back to normal, the propagation of crevice corrosion may continue at low temperature.

Cyclic dynamic polarisation tests are frequently used for testing the susceptibility to pitting and crevice corrosion. The potential is scanned anodically until the pitting or crevice initiation potential is reached and then reversed until repassivation at the pitting or crevice repassivation potential. For crevice corrosion testing, potentiostatic or stepwise potentiodynamic tests with relatively long holding times at each potential have to be applied. The chemical composition of the crevice electrolyte is then allowed to stabilise. In this way the critical potential of crevice corrosion initiation, E_{init}, and the critical potential of repassivation, E_{repass}, are established. Because of the complex nature of crevice corrosion, these potentials are not always as reproducible as they are for pitting corrosion. One problem is to control and reproduce the crevice geometry. Ranking of the most resistant stainless steels by use of E_{init} obtained at low temperatures is not a sensitive method since E_{init} is in the transpassive range. Thus, it is more common to rank high-alloyed stainless steels on the basis of the critical initiation temperature. In the same way that a repassivating potential for crevice corrosion exists, there may also be a repassivating temperature for crevice corrosion, T_{repass}.

In the first part of this paper, investigations on T_{init}, and T_{repass} for two high-alloy stainless steels are presented and the effects of potential and crevice-forming material are discussed.

In the case of stainless steels exposed to natural seawater a microbial slime layer will form on the surface. Recently, Holthe [4] published an extensive work on the effect that microbial activity has on the cathodic and the anodic properties of stainless steels exposed to natural seawater. While the biological activity does not influence the anodic properties, the cathodic properties of stainless steels exposed to natural seawater were found to change dramatically compared to artificial seawater or 3 % NaCl solution. The free corrosion potential rises from –100 mVSCE to above 300 mV SCE during 2–3 weeks of exposure. Values up to 450 mV SCE are measured after very long exposure. Furthermore, the microbial activity leads to a dramatic increase of the cathodic reaction rate after a few days of exposure. In the potential range from –400 to +200 mV SCE the cathodic reaction rate increases by two or three orders of magnitude compared to the sterile solutions. The effects seem to be independent of the stainless steel composition, but depend strongly on the seawater temperature. At temperatures above about 30°C, dependent on the microbiology in the seawater, the biofilm loses its activity [4, 5].

For conventional stainless steels, the increase of potential in natural seawater leads to pitting or crevice corrosion. This is usually not the case for high-alloy stainless steels. At temperatures below 30°C, the critical initiation potential is higher than +300 to +400 mV SCE and local corrosion will normally not initiate. On the other hand, the increase in cathodic efficiency is also important for high-alloy stainless steels. If crevice corrosion initiates at a temporary high temperature or in weak spots, the crevice corrosion rate increases dramatically compared to the rate in a 3 % NaCl-solution or in artificial seawater, and in galvanic couplings to less noble materials the corrosion rate of the less noble material can then be very high.

During propagation of crevice corrosion, the anodic reaction is mainly concentrated in the active crevice, and nearly all the cathodic reaction takes place on the passive area outside the crevice. Thus, the crevice corrosion rate may be determined by using well known mixed potential theory [3]. For a conventional stainless steel, the crevice corrosion rate usually depends greatly on the free surface area and the available cathodic reaction rate (cathodic control). For a high-alloy stainless steel, the anodic reaction rate may restrict the crevice corrosion rate for a lower ratio between cathodic to anodic area (anodic control). To study the effect of area ratio on the crevice corrosion rate, a wide range of experiments has to be performed. A suitable way to reduce the number of experiments is to combine experimental results with mathematical modelling.

The second part of this paper presents calculations of the crevice corrosion rate in pipe systems transporting natural seawater at 25 and 40°C by use of the computer programme GALVCORR. Overvoltage curves constructed on the basis of long time potentiostatic experiments, electrolyte resistance and tube geometry are the programme inputs.

2. Initiation and Repassivation of Crevice Corrosion

2.1 Experimental

The chemical composition of the two tested high-alloy stainless steels are given in Table 1. The specimens were machined from 3 mm thick rolled materials to rings with an inner diameter of 7 mm and outer diameter of 24 mm. A hole with a diameter of 1.3 mm was drilled near the outer circumference to suspend the sample on platinum wires. The surface preparation was as follows: automatic grinding on 220 SiC paper with carborundum paste 280 and glycol before grinding on a Petrodisc, diamond spray with grain size 15 μm. Thereafter, cleaning with soap, rinsing in ethanol and finally cleaning in acetone with ultrasound. Before mounting, the specimens were stored in desiccators with P_2O_5 drying agent for at least 3 days. The samples were mounted as shown in Fig. 1. Three different crevice formers were tested: metal/metal, metal/commercial gasket (Klinger Sil) and in most cases metal/polymer (POM). Total crevice

Table 1 Chemical composition of the alloys (wt%)

Material	C	Si	Mn	P	S	Cr	Ni	Mo	Cu	N
254 SMO	0.018	0.37	0.38	0.024	0.001	20.0	18.0	6.1	0.64	0.206
SAF 2507	0.021	0.23	0.31	0.015	0.001	24.8	7.15	3.86	0.05	0.296

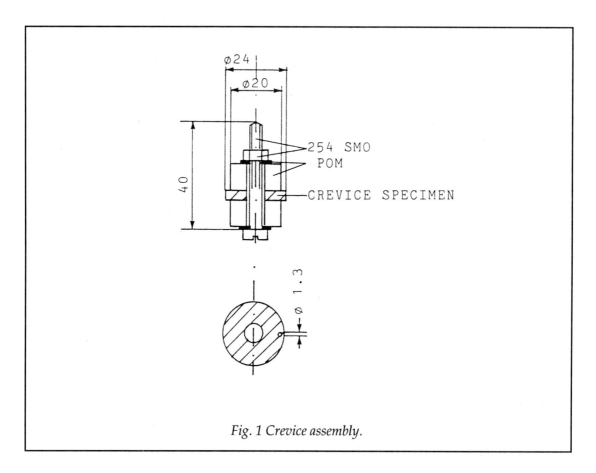

Fig. 1 Crevice assembly.

area under the gaskets was 5.5 cm². The different parts were mounted using a torque of 3 Nm.
Two different test procedures were used:

(a) Three specimens were suspended from on platinum wires and exposed to a 3 % NaCl solution in a 5 l glass beaker. The samples were freely exposed for 24 h at 20°C and then polarised by use of a potentiostat to 600 mV SCE. The temperature was raised in steps of 4 deg. each 24 h. When at least two of the three specimens had initiated crevice corrosion, the temperature was lowered to the actual temperature (32 or 40°C). Then, the potential was regulated to the actual level to measure the repassivation temperature. The temperature was further lowered in steps of 4 deg. each 24 h until repassivation. To limit the propagation of corrosion after initiation of crevice corrosion, a 1000 Ω resistance was used. At the temperature where the potential was regulated, the resistance was set to 10 Ω.

(b) Six specimens were freely exposed to a 3 % NaCl solution at 75°C in a 5 l glass beaker.

After 1 h the specimens were polarised to 300 mV SCE, which give immediate initiation of crevice corrosion. Simultaneously, the heater was switched off and the temperature was lowered to 40°C (12 h). Even here a 1000 Ω resistance was used to limit the propagation of crevice corrosion. After 24 h the temperature was lowered to 36°, the potential was regulated to the actual potential to measure repassivating temperature and the resistance was changed to 10 Ω. Thereafter, the temperature was lowered in steps of 4 deg. each 24 h. The experimental arrangement is schematically shown in Fig. 2.

In both procedures tap water was used as the cooling medium. Therefore, the lowest obtainable temperature was 10°C.

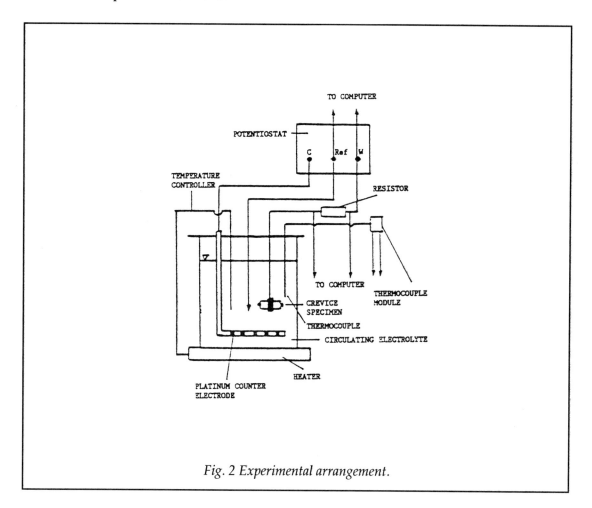

Fig. 2 Experimental arrangement.

2.2 Results

Some of the results from test procedure (a) are shown in Fig. 3. The temperature is plotted vs the last current at each temperature for crevice specimens of 254 SMO with plastic gaskets (POM). During initiation the potential was kept constant at 600 mV SCE and during repassivation at 0 mV SCE. T_{init} is defined as the temperature where the anodic current is stable at a value higher than 5.5 µA (1 µAcm^{-2}). Correspondingly, T_{repass} is defined as the temperature where the anodic current again is stable at a value lower than 5.5 µA (1 µA cm^{-2}). There is a large scatter in the results, especially in regard to the repassivation temperature. The scatter is probably caused by differences in the crevice geometry. Initiation of crevice corrosion was registered at different temperatures and time. Despite the restriction of anodic current after

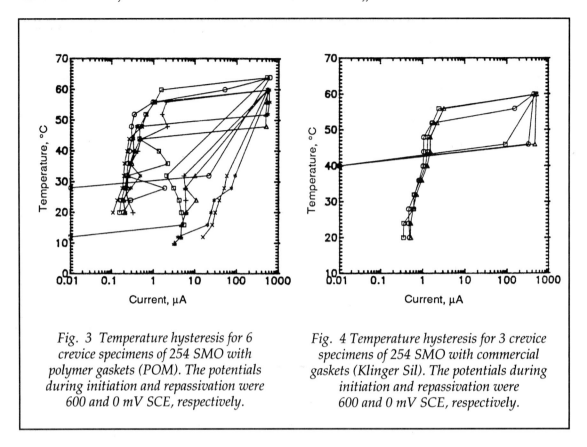

Fig. 3 Temperature hysteresis for 6 crevice specimens of 254 SMO with polymer gaskets (POM). The potentials during initiation and repassivation were 600 and 0 mV SCE, respectively.

Fig. 4 Temperature hysteresis for 3 crevice specimens of 254 SMO with commercial gaskets (Klinger Sil). The potentials during initiation and repassivation were 600 and 0 mV SCE, respectively.

initiation, corrosion has caused differences in crevice geometry. However, the results clearly show that the repassivation temperature may be considerably lower than the initiation temperature. While the initiation temperature is about 60°C, it was possible to maintain a low active corrosion after initiation at a temperature as low as 10°C both for 254 SMO and SAF 2507 (Fig. 5(b)) materials. Also metal/metal specimens were active at this low temperature.

A test with a commercial gasket (Klinger Sil) as a crevice former is shown in Fig. 4. For these specimens it was impossible to maintain active corrosion at 40°C or lower, while the initiation temperature is of the same order as for the other crevice formers.

Test procedure (b) gave a more uniform corrosion attack during the initiation phase and less scatter in the results compared to initiation by test procedure (a). Figure 5(a) and (b) show the results when the passivation during repassivation was 0 mV SCE. One of the six specimens of each material did not initiate, nor did the corrosion stop at an early stage. All other specimens were active at low temperature and four out of six were active after 24 h at 10°C both for rolled material of 254 SMO and SAF 2507.

T_{repass} is also measured by test procedure (b) for SAF 2507 at 300 and 600 mV. At 300 mV T_{repass} was 16°C for three specimens and 12°C for the other three. At 600 mV two specimens repassivated at 28°C and the other four at 24°C. These results are shown in Fig. 6 together with the results at 0 mV. There is no doubt that T_{repass} is increasing with increasing potential. This may be a surprising result, and an explanation is given in the following section.

2.3 Discussion

The above experimental results show that the repassivation temperature for the tested high-alloy stainless steels is much lower than the initiation temperature. In this respect there is a clear analogy with the initiation and the repassivation potentials observed in cyclic polarisa-

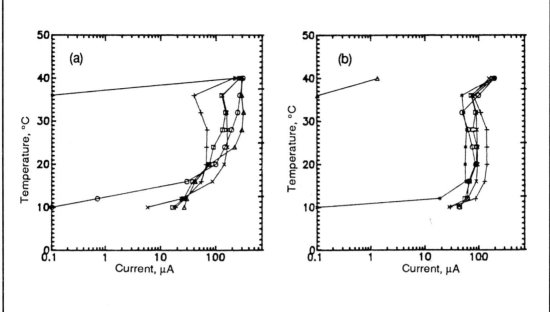

Fig. 5 Repassivating temperatures for (a) 6 crevice specimens of 254 SMO with polymer gaskets (POM), and (b) 6 crevice specimens of SAF 2507 with polymer gaskets (POM). The specimens are inititated at 300 mV SCE and 75°C, while the potential during repassivation was 0 mV SCE.

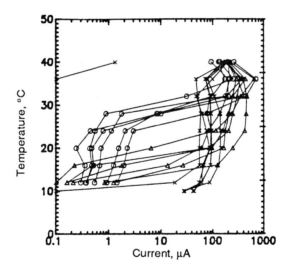

Fig. 6 Repassivating temperature for crevice specimens of SAF 2507 with polymer gaskets (POM). The specimens are initiated at 300 mV SCE and 75°C. The potentials during repassivation were (o) 600, (Δ) 300 and (×) 0 mV SCE.

tion tests. The reason for this hysteresis is largely the same in both cases—the possibility of maintaining a sufficient aggressive solution inside the occluded area.

Before initiation, the aggressiveness of the solution inside the crevice is determined by the passive current density and the geometry of the crevice. At low temperature the passive current density is quite low and the pH inside the crevice is above the critical value for the material, pH_{crit}. pH_{crit} is a function of the temperature — it increases with increasing temperature. As the temperature increases, so does the passive current density, which makes a more aggressive crevice solution. Thus, at a certain temperature the solution pH becomes lower than pH_{crit} and initiation occurs. Then the currents inside the crevice increase by orders of magnitude, which leads to a further lowering of pH due to hydrolysis. The large number of metal ions being dissolved attract chloride ions from the outside to maintain the charge balance, and the solution becomes saturated with metal chlorides. Precipitation of a very thin salt layer takes place until the ion flux is balanced with the diffusion flux out of the crevice. During steady state propagation, it is therefore a saturated chloride solution (9–10 M) and a very low pH (–1 and lower) inside the crevice.

When the temperature is lowered to obtain repassivation, it is the active corrosion current density and not the passive current density that determines the temperature for repassivation. The active corrosion current density is a function of several parameters, like the potential and the temperature, as indicated in Fig. 7. The curves in this Figure have been constructed from piecewise linear segments fitted to polarisation measurements in simulated crevice electrolytes [6–8]. This figure shows that with low outside potentials, e.g. 0 mV SCE, the inside potential will be in a potential range below 0 mV SCE where there is quite a large active corrosion current density in a near saturated solution (pH=–1.0), even at a temperature as low as 10°C. Thus, this active peak at low potentials explains the ability of the material to maintain an active crevice down to very low temperatures.

At higher outside potentials, e.g. 600 mV SCE, the potential inside has been measured to be in the range 0 to 300 mV SCE (results not shown here). According to Fig. 7 the current density in this potential region is very low when the temperature is down to 10°C. This indicates that repassivation should be expected at a higher temperature than at 0 mV SCE outside potentials, as we have indeed observed experimentally.

The crevice geometry is another factor which influences the repassivating temperature. In these experiments, the same crevice assembly was used in all tests and the currents were restricted at high temperature to get a reproducible crevice geometry and thereby reproducible results. In general, the repassivation temperature increases with increasing crevice width due to a higher diffusion flux out of the crevice. For the tests with a commercial gasket it was impossible to maintain active corrosion for temperatures below 40°C. Visual examination of these specimens after exposure showed corrosion attack at only a few points. The rest of the crevice area was covered with material from the gasket. It appears that the gasket is gluing itself to the metal surface which gives a very tight crevice and a little electrolyte volume. For such a tight crevice redistribution of ion concentrations and current density may not occur easily. Inspection of attack profiles of less tight crevices at various temperatures (not shown here) indicates that such a redistribution may be a prerequisite for maintaining an active crevice. This may then explain the very high repassivation temperature observed with the soft commercial gaskets. For surfaces facing commercial gaskets in practical systems, like large flange connections, the same tightness may not be present over the entire surface due to uneven surface pressure and a less smooth surface finish. Thus, crevice corrosion may also be possible with commercial gaskets at temperatures below 40°C.

Apart from being an interesting observation with respect to the mechanism of crevice corrosion, the temperature hysteresis also has a practical implication which can lead to serious corrosion problems. If a temporary temperature increase during service initiates crevice corrosion, the materials studied here, and others of similar quality, may remain active even if the temperature is brought back to a low level within a short time.

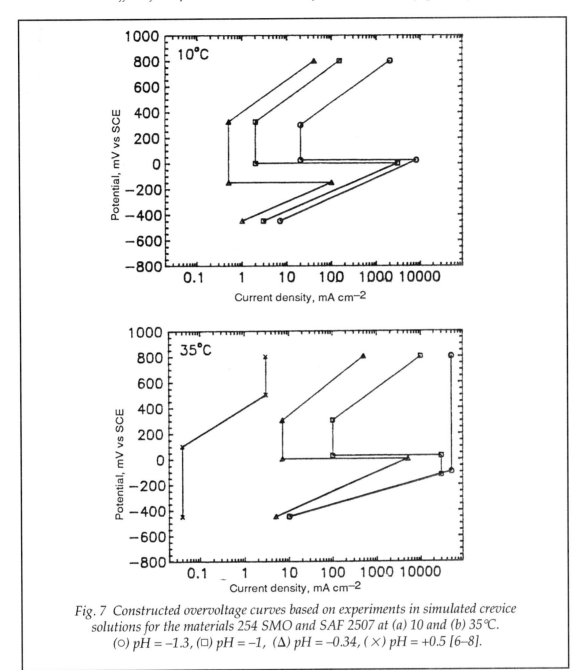

Fig. 7 Constructed overvoltage curves based on experiments in simulated crevice solutions for the materials 254 SMO and SAF 2507 at (a) 10 and (b) 35°C. (○) pH = –1.3, (□) pH = –1, (Δ) pH = –0.34, (×) pH = +0.5 [6–8].

3. Propagation of Crevice Corrosion

3.1 Model inputs

In GALVCORR the geometry is defined by dividing the tube into a number of segments. The segments may have different lengths and radii and they may be further divided into smaller elements. In this model study, tubes of three different radii—0.01, 0.025 and 0.15m —are used. The lengths of the tubes are varied from 0.002 to 1000 m. The active crevice is placed in the middle of the tube as a segment of its own. The defined geometry gives a total crevice area of 5.5 cm^2.

A pair of overvoltage curves, one anodic and one cathodic, is assigned to each segment. The curves used in this study are shown in Fig. 8. The active anodic curves are constructed on basis of long duration potentiostatic tests at 25° and 40°C with crevice specimens of 254 SMO [8]. The crevice assembly was of the same type as shown in Fig. 2 (POM gaskets). Already at this point it has to be stressed that the calculated results will be valid only for crevices of exactly the same type. Another important point is that the attacked areas tend to repassivate after some time of exposure. The longest time before repassivation observed was 59 days. The depth of a crevice corrosion attack will thus be restricted, and active crevice corrosion may not necessarily lead to practical corrosion failures.

The cathodic curve for temperatures below 30°C, reflects the effect of microbial activity. At 40°C there is no effect of the biofilm. Then the cathodic curve for temperatures above 30°C is used. The cathodic curve used for the active area has a lower cathodic reaction rate than the cathodic 40°C curve. The curve has no influence on the results and is not shown here.

The calculations are made by choosing a suitable start potential. Then the potential and current distribution along the tube are calculated by linearization of the non-linear boundary conditions and an iterative solution of the unknown potentials.

The resistivity of seawater is set to 20 ohm cm at 25°C and to 15 ohm cm at 40°C. To calculate the corrosion rate in mm/year, some material data are necessary inputs.

3.2 Model results

The potential variations for different radii and lengths are given in Fig. 9 (a)–(c) for the calculations at 25°C and in Fig. 10 (a)–(c) for the calculations at 40°C. The active crevice placed in the middle of the tube has the lowest potential. The trend is less for the tube with the largest radius since a greater electrolyte volume combined with the high conductivity of seawater, leads to a low potential drop.

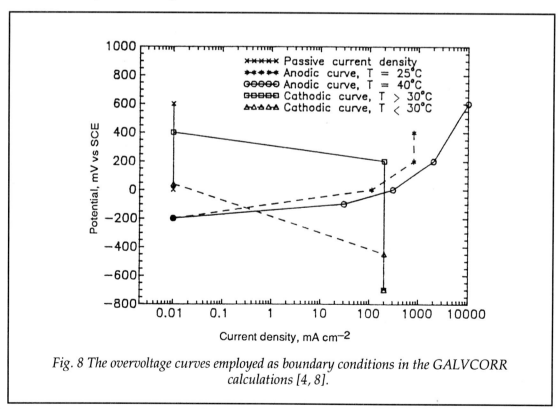

Fig. 8 The overvoltage curves employed as boundary conditions in the GALVCORR calculations [4, 8].

Figure 9(d) (25°C) and Fig.10(d) (40°C) show the calculated corrosion rates as a function of pipe geometry. The results are quite different at the two temperatures. This is discussed further in the next section.

3.3 Discussion

At 25°C the potential increases when the length increases up to a maximum potential around 350 mV SCE. The potential also increases with increased radius when the length is maintained constant. This indicates that the effective cathode current becomes larger for an increase in radius and length. When the potential reaches values above 300 mV the cathodic reaction rate is so low that current from the remaining part of the tube is negligible compared to the total galvanic current. However, this is unimportant for the results at 25°C. The anodic overvoltage curve at this temperature shows that when the potential outside the crevice becomes 200 mV or higher, the corrosion rate will not increase any further. This is also illustrated in Fig. 9(d) where the corrosion rate reaches the upper plateau for very short tube lengths. Even for the smallest tube, the maximum corrosion rate is obtained for a length of *ca.* 0.035 m. The ratio between the cathodic to anodic area then is 4. This shows that at 25°C in natural seawater with microbial activity, the anodic properties may restrict the corrosion rate (anodic control) at a very low ratio between the cathodic to anodic area.

Due to the low cathodic reaction rate at 40°C, the calculated results here are totally different to those at 25°C. The maximum corrosion rate is obtained when the potential outside the crevice reaches the free corrosion potential. Figure 10(d) shows that for the tube of largest radius, the maximum corrosion rate is not achieved even for a length of 1000 m and it is the cathodic reaction rate which restricts the corrosion rate (cathodic control). The ratio between the cathodic to anodic area then is higher than 1.7×10^6. Figure 10(d) also shows that the corrosion rate increases with increased length until there is no more cathodic capacity left. The effective length is about 40 m for the tube of radius 0.01 m, about 100 m for the tube of radius 0.025 m, and larger than 1000 m for the tube of radius 0.15 m. The maximum corrosion rate decreases with reduced radius. This is due to the higher ohmic potential drop along the tube when the radius reduces.

The calculations clearly demonstrate the effect microbial activity has on the crevice corrosion rate. To obtain the maximum corrosion rate only a small cathodic area is necessary at 25°, while the cathodic area at 40°C has to be very large.

At a constant potential the corrosion rate at 40°C is higher than at 25°C. However, the low cathodic capacity at temperatures above 30°C causes a lower corrosion rate at the highest temperature. As an illustration, the maximum corrosion rate at 25°C (0.9 mm/year) is obtained for a length of 0.035 m (R = 0.01). Correspondingly, the corrosion rate at 40°C for the same geometry (L = 0.035 m, R = 0.01 m) is 0.002 mm/year, a factor of 450 in difference.

The present calculations have been performed with an anodic area of 5.5 cm^2. On a pipe system, if there is more than one attack with a total crevice area greater than 5.5 cm^2, the average corrosion rate will be lower. On the other hand, if there is only one single attack with a crevice area considerable lower than 5.5 cm^2, the corrosion rate may be higher.

It also has to be remembered that the anodic curves are based on experiments where the circular crevice assembly with POM gaskets shown in Fig. 2, is used. The expected corrosion rate for metal/metal specimens will be in the same range as for the assembly with POM gaskets [8]. However, as shown in Fig. 4, it was impossible to maintain active corrosion when the gaskets were of a commercial type (Klinger Sil). Anodic curves based on other materials than the tested 254 SMO may also be different from the curves given in Fig. 8. It is also important that the crevice corrosion attacks tend to repassivate. All specimens tested were repassivated before two months of exposure.

The results at 40°C show that the available cathodic current restricts the corrosion rate. The

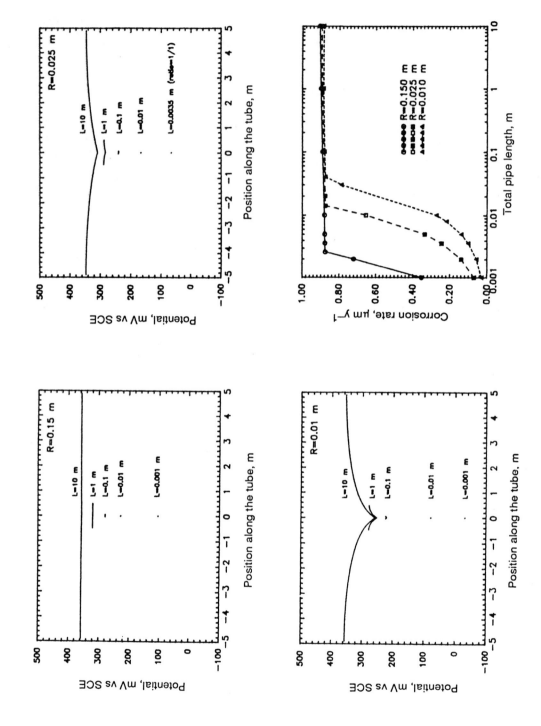

Fig. 9(a–c) Calculated potential profiles in natural seawater at 25°C for different pipe geometries. An active crevice is placed in the middle of the tube. (d) Calculated crevice corrosion rate of 254 SMO in natural seawater at 25°C as a function of pipe geometry.

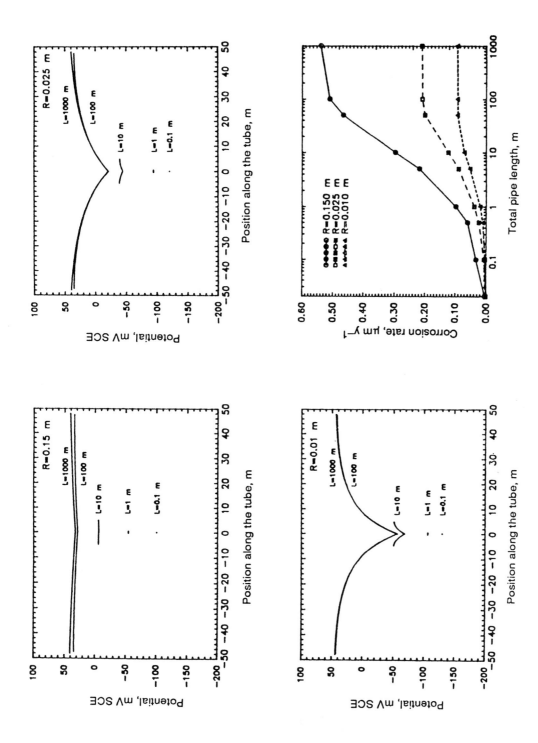

Fig. 10 (a–c) Calculated potential profiles in natural seawater at 40°C for different pipe geometries. An active crevice is placed in the middle of the tube. (d) Calculated crevice corrosion rate of 254 SMO in natural seawater at 40°C as a function of pipe geometry.

rate increases with increased length up to a limit. The limit is not reached for a tube of 1000 m length and radius 0.15 m, which gives an enormous ratio between the cathodic to anodic area. Thus, it is impossible to obtain realistic corrosion rates in a laboratory experiment when the galvanic coupling concept is used. To combine separate anodic and cathodic curves together with mathematical modelling is therefore a better method when predicting corrosion rates. To use the galvanic coupling concept in a laboratory, is no problem in natural seawater at 25°C, where a relative small area ratio is necessary.

Due to the low cathodic current at 40°C, a weak corrosion attack will decrease the corrosion potential over a relatively long distance. This may reduce the probability for initiating new sites of attack over the same distance. However, the intensity of the first initiated attack may then rise with time.

4. Conclusions

1. Many high-alloy stainless steels are quite resistant to crevice corrosion at temperatures up to 30°C. In this study, experimental evidence is given to show that crevice corrosion may propagate at temperatures far below 30°C, provided corrosion is initiated at a higher temperature.

2. A lower temperature limit exists for propagation of crevice corrosion of high-alloy stainless steels (the repassivating temperature).

3. There seems to be a clear analogy between the temperature and the potential concerning the hysteresis associated with the initiation and the repassivation values.

4. The repassivating temperature depends on the crevice forming material. Gaskets made of a polymer material (POM) and metal/metal specimens show repassivating values from 10° to 25°C, while the use of a commercial gasket (Klinger Sil), made it impossible to maintain active corrosion at 40°C or lower.

5. The repassivating temperature also depends on the potential. With metal/metal or metal/polymer crevices the tested specimens repassivated around 10°C at a potential of 0 mV SCE. At 300 and 600 mV SCE the average values were 14° and 25°C, respectively.

6. The computer programme GALVCORR is used to calculate the crevice corrosion rate and to demonstrate the effect microbial activity has on the cathodic reaction rate in natural seawater, and thereby on the corrosion rate for an active crevice within pipelines of different lengths and radii. The calculations are based on long time potentiostatic experiments at 25° and 40°C with small specimens. The calculated corrosion rate at 25°C is 2 to 3 decades higher than the corrosion rate at 40°C when the cathodic area is constant.

7. Due to the high cathodic reaction rate at 25°C, a very small cathodic area is necessary to obtain the maximum corrosion rate and the corrosion rate is restricted by the anodic reaction rate.

8. In contrast, the corrosion rate is restricted by the available cathodic area due to the low cathodic reaction rate at 40°C. To obtain the maximum corrosion rate, it is necessary to have extremely large cathodic areas. Thus, in the absence of an active biofilm, it is impossible to obtain realistic corrosion rates in a laboratory test with specimens of limited size.

5. Acknowledgement

The authors gratefully acknowledge the financial support given by The Norwegian Council for Scientific and Technical Research (NTNF), The Norwegian Petroleum Directorate and the industry companies Avesta, Sandvik Steel, Statoil, Stavanger Staal and Framo Engineering.

References

1. P. O. Gartland, GALVCORR 2.0, User's manual, SINTEF report STF16 F86126, 1986.
2. R. Johnsen, Stainless steels in seawater systems, NITO Conference, The Norwegian Society of Engineers, in cooperation with Nickel Development Institute, Amsterdam, 7–8 February, 1990.
3. J. M. Drugli and E. Bardal, A Short Duration Test Method for Prediction of Crevice Corrosion Rates on Stainless Steels, Corrosion, 1978, **34**, 12.
4. R. Holthe, The cathodic and anodic properties of stainless steels in seawater, Dr. thesis, Norwegian Institute of Technology, 1988.
5. A. Mollica *et al.*, Cathodic performance of stainless steels in natural seawater as a function of microorganism settlement and temperature, Corrosion, 1989, **45**, 1.
6. P. O. Gartland and S. Valen, Initiation of crevice corrosion of high-alloyed stainless steels in chlorinated seawater, SINTEF report STF34 F90059, 1990 (in Norwegian).
7. P. O Gartland and S. Valen, Crevice corrosion of high-alloyed stainless steels in chlorinated seawater— II, Aspects of the mechanism, Corrosion NACE '91, Cincinnati, OH, paper 511.
8. S. Valen, Initiation, propagation and repassivation of crevice corrosion of high-alloyed stainless steels in seawater, Dr. thesis, Norwegian Institute of Technology, 1991.

An Intelligent Probe for *in situ* Assessment of the Susceptibility of Hydrogen Induced Cracking of Steel for Offshore Platform Joints

W. WEI, D. PENG*, F. CHAO, L. ZHENG AND D. YUAN-LONG

Institute of Corrosion and Protection of Metals, Chinese Academy of Sciences, Shenyang 110015, China
* Shenyang Institute of Automation, Chinese Academy of Sciences, Shenyang 110003, China

Abstract

In view of the synergistic effect of the local pollution of the seawater next to the surface of the offshore platform joints and cathodic protection, an intelligent probe technique has been developed for assessing the susceptibility to hydrogen induced cracking. The method is based on the measurement of the permeation rate of atomic hydrogen in a sensor and uses a microcomputer system with specially designed software to provide *in situ* evaluations. In this paper, the design, structure and typical results are presented and discussed.

1. Introduction

The susceptibility of steel for offshore platform joints to hydrogen induced cracking (HIC) will be greatly increased as a result of the following effects often operating synergistically:

(i) Seawater and mud at an estuary or harbour are often polluted [1] and the seawater on the sheltered zone underneath the fouling organisms and next to the surface of the joints may also be locally polluted with hydrogen sulphide [2];
(ii) loading, residual and stochastic stresses are always concentrated on the platform joints and often reach rather high levels;
(iii) in line with the conventional point of view held by many designers of offshore platforms, joints are often protected cathodically at very negative potentials, provided that hydrogen evolution can be avoided [3];
(iv) steels for offshore platform joints may be sensitive to HIC even without cathodic protection in seawater [4].

In practice, there is no commercially available probe technique for *in situ* measuring and assessing HIC in steel joints in clean and in polluted seawater.

For these reasons, a new intelligent probe technique has been developed for safety inspection and control of offshore platforms [5, 6]. In this paper, the probe design, structure and the underwater application are presented.

2. Design and Structure

The probe assembly, consists of a 'gun-shaped' probe and a periphery attachment box. In the probe, a microcomputer system (MCS) is mounted in a watertight compartment. At the front of the 'gun' probe, a hermetically sealed fuel cell type sensor of hydrogen-metallic oxide [7, 8] with a button shape is mounted for measuring the permeation rate of atomic hydrogen (Fig.1).

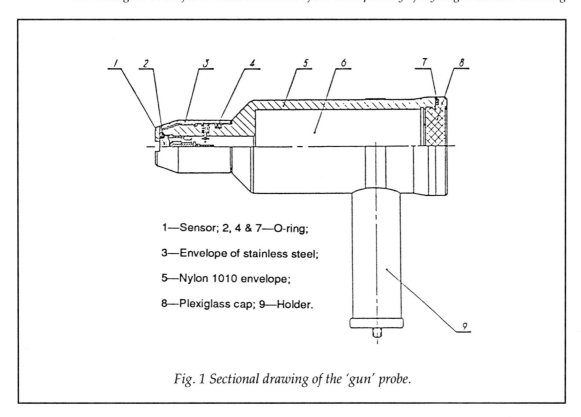

1—Sensor; 2, 4 & 7—O-ring;
3—Envelope of stainless steel;
5—Nylon 1010 envelope;
8—Plexiglass cap; 9—Holder.

Fig. 1 Sectional drawing of the 'gun' probe.

When the 'gun' probe is being operated, the front of the probe is inserted and mounted into the covered zone next to the surface of the joint in the seawater. After one hour, the 'gun' probe is recovered from the seawater by a diver. It is then connected to the periphery attachment box. By using specially designed software the relative permeation rate of atomic hydrogen in steel and the HIC susceptibility of the joint steel is printed out on an ultramicroprinter. The schematic diagram of the system is shown in Fig. 2. The software is designed on the basis of the correlation that exists between the level of the output signals of the sensor and the strength/embrittlement of the joint steel—as established in other work[9].

In the operating condition of the 'gun' probe in clean or polluted seawater, a constant current is applied to the outer surface of the sensor for cathodic charging of hydrogen to increase the signal level and decrease the corrosion of the sensor surface. The short-circuited current of the sensor—the relative permeation rate of atomic hydrogen in steel—will input into the signal receiving system of the MCS. The MCS is designed for data sampling, logging, storing, processing, result judgement and controlling the data output.

The attachment box, ultramicroprinter, charging power supply for Cd–Ni battery and the button for controlling and selecting output are mounted in the periphery.

The output form to print the results is shown in Table 1. The designed pressure resistance of the 'gun' probe for underwater use is 1.5MPa and the designed safety coefficient is 2.5.

The fundamental characteristic parameters and specifications of the probe are as follows:

- Sensor — sealed button-shaped fuel cell-type sensor of hydrogen–metallic oxide;
- Measuring range of the current — 0–250 µA (or higher, if needed);
- The results of the evaluation — 'Safe', 'Dangerous' or 'Sensitive to failure hazard (H.F.), of the joint to HIC respectively in the corresponding media and the protection potential ranges;
- Input impedance of the MCS — less than 0.3 Ω;

- Memory capacity — 96 bytes for stored data;
- Precision of the MCS — higher than 0.5%;
- Pressure-bearing capacity — at least 1.0 MPa(or higher, if needed);
- Time for underwater use for each charging of the Cd–Ni battery of the probe — 2 h;
- Weight of the 'gun' probe in the air — approximately 1.5 Kg or so;
- No cables required under sea water;
- Portable.

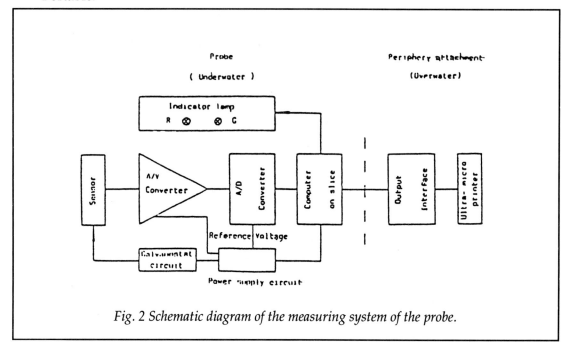

Fig. 2 Schematic diagram of the measuring system of the probe.

Table 1 The output form of the inspection results in store in MCS

The permeation rate of atomic hydrogen and the hazard of hydrogen induced cracking of steel for platform joint
No. ct.   No. ct.   No. ct.   No. ct.
Point No. xx
Max. current: xx uA

(1) For general corrosion region:
* In clean seawater, or
* In slight H_2S polluted seawater, or
* In moderate H_2S polluted seawater, or
* In heavy H_2S polluted seawater
To be tabulated by one set of the results according to the current value measured.
More positive than xxx mV: Safe
Between xxx & xxx mV: Dangerous
Less positive than xxx mV: H.F.

(2) For localised corrosion region:
Less positive than 300 mV: H.F.
*Potential tabulated is vs Zn electrode

---

3. Results and Discussion

(i) *Pressure-resistance test*: This is carried out in an autoclave with artificial seawater under a pressure of 1.0 MPa for two hours. The results show that the probe can be used at a depth equivalent to at least 100m of seawater.

(ii) *Laboratory test*: Tables 2 and 3 show the results of the probe respectively dipped into clean seawater and seawater polluted with hydrogen sulphide (less than 20ppm). The results coincide with the correlation between the output signals of the sensor and the strength/embrittlement of the joint steel calibrated in other work [9].

(iii) *Underwater uses of safety inspection of offshore platform joints*: The probe has been successfully used in the safety inspection on four offshore platforms, in the Bohai Sea and in the Nanhai Sea, China. A typical inspection result taken by this probe from No.133 joint of U/A-B platform at a depth of 15.24m is shown in Table 4. The results taken from the above offshore platforms by this probe coincide with the actual phenomena obtained from physical inspection methods, e.g. underwater *in situ* magnetic particle testing. In a new offshore platform operated for only five years in seawater, cracks have been found on almost all the welding seams of the joints that have been inspected by these two methods, although the design was according to international criteria and protection was established using conventional criteria. It is obvious that some new information can be provided by this probe for the safety inspection of offshore platform joints in the safety inspection of offshore platform joints in the actual media next to the surface of the joints. The results also revealed that, there are some further aspects to be considered over and above the conventional criteria used in the design and structure, strength and cathodic protection aspects of offshore platforms.

Table 2 Results of lab test in artificial seawater

The permeation rate of atomic hydrogen and the hazard of hydrogen induced cracking of steel for platform joint

No.	ct	No.	ct	No.	ct	No.	ct
01	00	02	00	03	00	04	00
05	00	06	00	07	00	08	00
45	00	46	00	47	00	48	00
49	00	50	00	51	00		

Point No. 51
Max. current: 00 μA
Status:
(1) For general corrosion region:
* In nil H_2S polluted seawater
More positive than 90mV: Safe
Between 90 & 50 mV: Dangerous
Less positive than 50 mV: H. F.

(2) For localised corrosion region:
Less positive than 300 mV: H. F.

* Potential tabulated is vs Zn electrode

Table 3 Results of lab test in artificial seawater with H_2S (less than 20 ppm)

The permeation rate of atomic hydrogen and the hazard of hydrogren induced cracking of steel for platform joint

	No.	ct	No.	ct.	No.	ct.	No.	ct.
	01	00	02	00	03	00	04	01
	05	02	06	04	07	05	08	06
	53	16	54	16	55	16	56	17
	57	17	58	16	59	16	60	16

Point No: 57
Max. current: 17μA
Status:
(1) For general corrosion region:
* In slight H_2S polluted seawater
More postive than 110 mV: Safe
Between 110 & 80 mV: H. F.

(2) For localised corrosion region:
Less positive than 300 mV: H. F.
*Potential tabulated is vs Zn electrode

Table 4 Underwater test results on No. 133 joint of U/A-B platform

Depth: –15.246 m; Sensor No: MR9–249
The permeation rate of atomic hydrogen and the hazard of hydrogen induced cracking of steel for platform joints

	No.	ct.	No.	ct.	No.	ct.	No.	ct.
	01	00	02	00	03	00	04	00
	05	00	06	00	07	01	08	02
	69	09	70	09	71	09	72	09
	73	09	74	10	75	09		

Point No: 74
Max. current: 10μA
Status:
(1) For general corrosion region:
* In slight H_2S polluted seawater
More positive than 110 mV: Safe
Between 110 & 80 mV: Dangerous
Less positive than 80 mV: H. F.

(2) For localised corrosion region:
Less positive than 300 mV: H. F.
*Potential tabulated is vs Zn electrode

4. Conclusions

The following conclusions can be drawn based on simulation tests in the laboratory and on safety inspections of the offshore platforms: on U/A-B offshore platform; Zengbei Oil Field, JZ 20-2-1, JZ 20-2-3 and Bohai No.8 offshore platform; Bohai Sea, and WZ 10-3 offshore platform, Nanhai Sea, China:

(1) The probe is a new intelligent instrument device for safety inspection, specially designed and developed for offshore platform for *in situ* inspection, assessment and judgement of the HIC susceptibility of the steel of offshore platform joints under cathodic protection in seawater.

(2) The relative permeation rate of the atomic hydrogen in steel, and a comprehensive evaluation of the HIC susceptibility of the joints can be printed and tabulated out in clean or hydrogen sulphide-containing polluted seawater, including that locally polluted with hydrogen sulphide next to the surface of the joints in seawater or mud. Prospects for further applications are encouraged by the successful inspection of the probe on five of the above offshore platforms.

References

1. Chemical Engineering Progress Symposium Series, 1969, No. 97, 'Water', 1968, Vol. 65; or cf., Translating Collection of Geological Environmental Pollution — Water Pollution, 9, Scientific & Technical Reference Publishing House, 1973.
2. D. Yuan-long *et al.*, Technical paper collection of the studies on the intelligent probe technique for *in situ* assessing the susceptibility of HIC of the steel for offshore platform, unpublished report, 1988.
3. Control of corrosion on steel, fixed offshore platform associated with petroleum production.Recommended practice, NACE Standard RP-01-76, 1976.
4. W. Zhounguang, Z. Zuming and P. Jun, Acoustic emission monitoring of fatigue crack growth in SM50B-ZC steel used in the structure of offshore platform, Proc. Int. Conf. on Evaluation of Materials Performance in Severe Environments, Kobe, Japan, Nov, 1989.
5. D. Yuan-long *et al.*, Chinese invention Pat: 89105157.0.
6. D. Yuan-long *et al.*, Technical paper collection of the studies on the electrochemical techniques for *in situ* inspection of offshore engineering structures, unpublished report, 1988.
7. D. Yuan-long *et al.*, Chinese Invention Pat: 90106448.3.
8. D. Yuan-long *et al.*, Electrochemical probe for measuring the activity of HIC for offshore engineering structures, Proc. 1986 Conf. on Inspection, Repair and Maintenance for the Offshore and Marine Industries, Singapore, Jan., 1986.
9. D. Yuan-long *et al.*, Technical paper. Collection of the studies on the rationality of the criteria of cathodic protection of the steels for offshore platforms. Unpublished report, 1987.

12

Aspects of Testing Stainless Steels for Seawater Applications

P.O. GARTLAND

SINTEF, Trondheim, Norway

Abstract

About one decade of intensive research at SINTEF Corrosion Center on the local corrosion susceptibility of stainless steels in seawater is reviewed. In natural seawater it is shown that the formation of a biofilm on the surface has an important influence on the local corrosion initiation tendency and propagation rate for temperatures up to 32°C. In chlorinated seawater the tendency to local corrosion initiation is even stronger, but the propagation rate is typically less than in natural seawater. Surface preparation and ageing of the surface is shown to have a distinct influence on the initiation tendency, and this observation may be utilised during the start-up of a chlorinated seawater handling system in order to minimise the risk of local corrosion. Finally, it is discussed how short duration tests can be combined with results from exposure tests to establish a design curve valid for a wide range of seawater-like solutions.

1. Introduction

Testing of stainless steel for seawater applications may involve different types of tests, ranging from long term immersion tests in natural seawater to short time tests in more artificial electrolytes. When properly used, most of the tests will reveal significant information about the corrosion resistance of the steels in seawater.

The most important type of corrosion on stainless steels in seawater is local corrosion, in particular crevice corrosion and pitting at welds. Over the last decade much work has been carried out at SINTEF Corrosion Center related to these types of corrosion attack. Some of the investigations have been rather basic in nature, focusing on the electrochemical behaviour of an oxidised stainless steel surface in various environments like natural seawater, chlorinated seawater and simulated crevice solutions, while other studies have been more practical, trying to establish a better knowledge on how to avoid local corrosion in real systems [1–14].

The present paper highlights some aspects of testing stainless steels for seawater applications. The paper is divided into three parts. In the first part some of the findings in natural seawater are discussed, focusing on the cathodic effect of a biofilm. The second part looks at the behaviour of stainless steels in chlorinated seawater, which is known as an even more corrosive medium than natural seawater. The third section discusses short term testing, where the natural potential rise experienced in natural or chlorinated seawater is replaced by potentiostatic control.

1.1 Exposure test in natural seawater

When a stainless steel surface is exposed to natural seawater at moderate temperatures the corrosion potential is observed to rise with the time, as shown in Fig. 1. At temperatures below 32°C the potential rises up to about 300 mV SCE within 3–12 days, depending upon the temperature. Prolonged exposure has shown that the potential may continue to rise slowly up

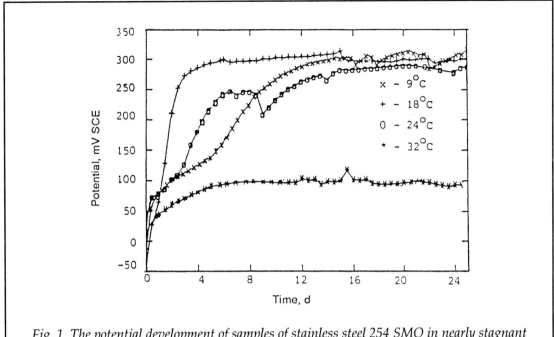

Fig. 1 The potential development of samples of stainless steel 254 SMO in nearly stagnant seawater, at four different temperatures.

to *ca.* 450 mV SCE after more than one year [2]. This potential rise has been observed in many different laboratories [15–20], and, is in all cases, attributed to the formation of a biofilm on the surface.

The marked difference in the potential level at 32°C and at lower temperatures indicate that the biofilm effect is absent at 32°C. This is further supported by the results in Fig. 2, where no biomaterial is present. In this figure, there is a small potential rise and a temperature effect, but nothing like that of Fig. 1 at low temperatures. The results of Fig. 2 can be understood from other observations, showing that the passive current density increases with the temperature [10] and reduces with time [5].

The potential rise in the presence of a biofilm has been observed on many different qualities of stainless steels, nickel alloys and also on titanium [3, 5]. This generality indicates that the effect is connected with changes of the cathodic reaction rather than the anodic reaction. This implies that most materials in the passive state will experience a similar potential rise. Copper alloys are an exception here [3], possibly due to the poisoning effect of the copper ions.

The practical consequence of the bioactivity is that passive materials at moderate temperatures will be subject to an increased risk of local corrosion initiation due to the potential rise. High alloy stainless steels like 254 SMO do not normally corrode in seawater at temperatures up to 30°C, but the more conventional stainless steels, AISI 304 or 316, are not resistant enough to withstand a potential rise up to 300–450 mV SCE, even in rather cold seawater.

The existence of a maximum temperature for the biofilm formation has been observed also by Scotto *et al.* [21]. However, the maximum temperature in the Mediterranean water seemed to be somewhat higher than 32°C. This can be attributed to the higher tolerance limit of the bacteria in the generally warmer seawater at the Italian coast. The results by Scotto *et al.* also indicated that the biofilm effect had a maximum limit of about 2.5 ms^{-1} with respect to the seawater flowrate. Our investigations have revealed no such limit in the range of flowrates up to 3.8 ms^{-1} [5].

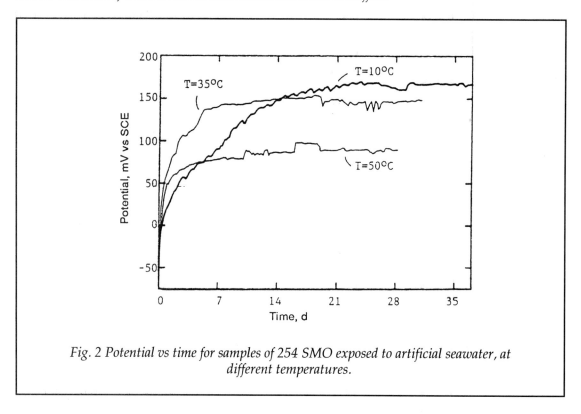

Fig. 2 Potential vs time for samples of 254 SMO exposed to artificial seawater, at different temperatures.

The interpretation of the biofilm effect as a cathodic effect mentioned above has not been based on the measurements of the potential rise alone. Cathodic polarisation measurements on plate samples, both potentiodynamic and potentiostatic, support this; Fig. 3 shows some typical results obtained with relatively slow potentiodynamic polarisation in natural seawater. The curve (marked 4 days, E_{corr}) is a typical polarisation curve in the absence of a biofilm, as similar curves are obtained in artificial seawater. The curve is characterised by a Tafel type behaviour down to about –500 mV SCE, below which there is a limiting oxygen current density. After 18 days exposure, i.e. when the potential has increased and levelled out, there is only a small change in the polarisation curve at the smallest current densities, where the curve bends upwards. This extra contribution can explain the potential rise realising that the corrosion potential is a mixed potential due to the cathodic reaction and an anodic reaction which here is a passive current density of the order 0.01–0.001 µA cm^{-2} [5].

What is particularly interesting with the results of Fig. 3 is the more dramatic changes of the cathodic properties following a cathodic polarisation to 100 mV SCE. We now observe that the previous small reaction rate at the upper potentials increases by several decades, and after a couple of weeks the whole cathodic curve is almost vertical and little potential dependent. The effect is also sensitive to the temperature in the same way as the potential rise, as shown in Fig. 4. This strongly supports the idea that the potential rise and the cathodic reaction increase at the upper potentials are both due to the same biofilm related mechanism. The details of the mechanism are not clearly understood, but some ideas have been discussed in a previous paper [4].

From a practical point of view, the cathodic reaction enhancement has serious implications. If local corrosion occurs, the potential will normally drop to a lower level until the available cathodic current balances the anodic current of the local pit or crevice. A large cathodic efficiency, as observed in the presence of a biofilm, means that a rather low area ratio between a cathode and an active anode is sufficient to allow rather large corrosion rates. This is

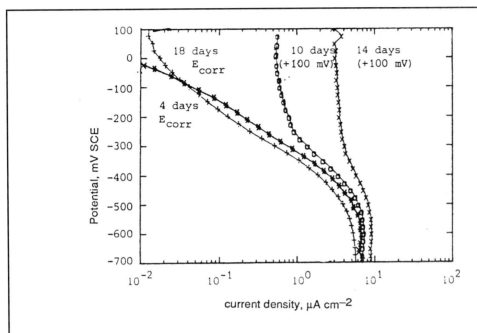

Fig. 3 Potentiodynamic curves for 254 SMO steel recorded after 1 and 18 d of free exposure and at 10 and 14 d at + 100 mV SCE respectively. Nearly stagnant seawater; temperature 18 ± 1°C.

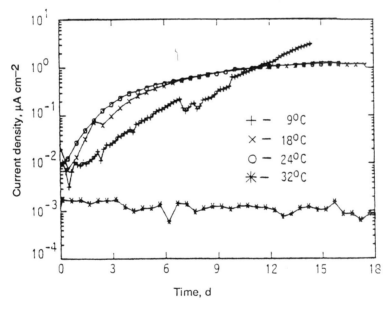

Fig. 4 The development of the cathodic cd on stainless steel 254 SMO at +100 mV SCE (0 mV at 32°C). Pre-exposed for 14 d at the free corrosion potential. Nearly stagnant conditions and 4 different temperatures.

demonstrated in Fig. 5, where actively corroding crevice samples of AISI 316 were coupled to an outer cathode. With area ratios as low as 5:1–100:1, the corroding crevices in natural seawater experienced an increase in the corrosion current up to 100–300 μA cm^{-2} or 1–3 mm/

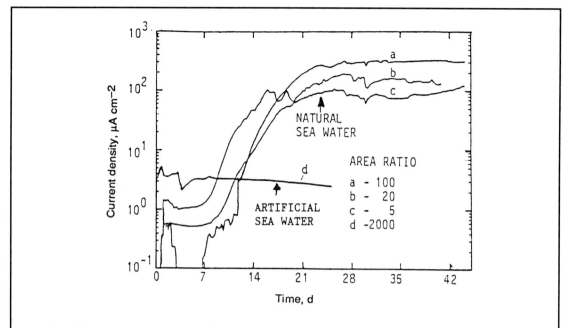

Fig. 5 Crevice corrosion rate of AISI 316L stainless steel in nearly stagnant natural and synthetic seawater at different ratios between cathode and anode area. Temperature 8 ± 1°C.

year after the initial 7–10 d time lag when the biofilm was being formed. In contrast, the crevice exposed in artificial seawater remained at a corrosion rate about two decades lower, even with the much larger area ratio of 2000:1.

A similar development of the corrosion rate was seen when a small 90/10 CuNi anode was coupled to a larger stainless steel cathode in seawater (Fig. 6).

The biofilm formed on the metal surface in natural seawater is due to the settlement of bacteria [6]. Thus, the biofilm effect should be considered as an example of MIC (microbial influenced corrosion). In the literature MIC is a subject of great concern, and a number of methods are utilised to characterise MIC, as recently reviewed by Mansfeld and Little [22]. In natural seawater there seems to be little doubt that the main effect of the bacterial activity is on the enhancement of the cathodic reaction. It is, therefore, surprising that measurements of cathodic reaction rates have not been utilised to a greater extent. In our opinion, it is one of the most sensitive indicators of the bio-activity relating directly to the local corrosion susceptibility.

1.2 Exposure tests in chlorinated seawater

Chlorine is added to seawater to kill living organisms, including bacteria. In this way the biofilm effects discussed above will not occur. However, the chlorine causes a series of new cathodic reactions, some of which have a very high equilibrium potential. The result is that the non-corroding stainless steel is raised to even higher potentials than in natural seawater. The actual potential level is a function of the chlorine concentration and the temperature, as shown in Fig. 7. A typical concentration for practical use is 0.5–1.0 ppm, and typical operating temperatures are 20–30°C. According to Fig. 7 we should then expect potentials in the range 600–650 mV SCE. It is no surprise, therefore, that local corrosion is initiated more easily in chlorinated seawater than in natural seawater. The conventional stainless steels have no chance to resist these high potentials even in cold seawater. The more highly alloyed steels, such as 6-Mo steels and the so-called superduplex steels can still be used, but at a somewhat

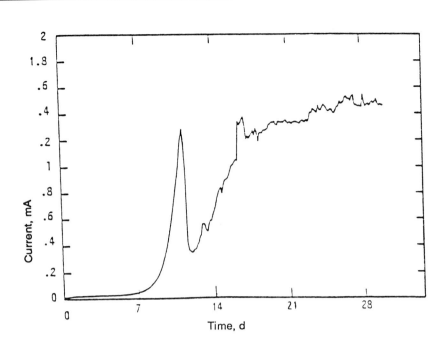

Fig. 6 Development of galvanic current between 90/10 CuNi (anode) and 254 SMO (cathode) at flow rate 1 m s^{-1} and temperature of 12°C. Anode area 10 cm^2, cathode/anode area ratio 45:1.

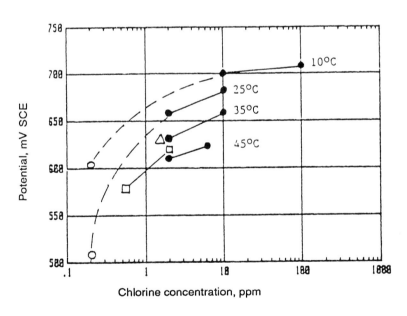

Fig. 7 Near steady state potentials measures on non-corroding samples of 254 SMO at different levels of chlorination and temperature.
● *Ref. [7], 14 d;* ○ *Ref. [6], 50 d;* □ *Ref. [24], 45°C, 90 d;* △ *Ref. [24], 35°C, 90 d.*

lower temperature than in natural seawater. Statoil [23] has recommended 30°C as the upper temperature limit for 254 SMO in chlorinated seawater.

The results of Fig. 7 were, in part, obtained by a series of experiments designed to compare the corrosion resistance of SAF 2507 and 254 SMO. The experimental arrangement is illustrated in Fig. 8. Crevice samples of the actual materials were immersed in heated, chlorinated seawater and coupled to large cathode plates of the same material. The potential of the whole assembly, as well as the current flowing between the crevice sample and the cathode, were monitored. Initiation of corrosion was observed from the current readings without having to remove the samples from the seawater. The test series was run at six different conditions, each of two weeks duration. The six conditions were nominally 25, 35 and 45°C, with two chlorine levels (2 and 10 ppm) at each temperature. Two types of samples were used. The 'old samples' were never taken out of the seawater but were exposed to all six conditions of gradually increasing severity. The 'new' samples were used for only one condition, i.e. for each new condition these samples had not previously been exposed. The testing was carried out in triplicate at each condition. Further details are given elsewhere [7]. The results were not particularly surprising in view of the behaviour of the 'new' samples of both materials. Figure 9 shows that the new samples of 254 SMO resisted 35°C and 2 ppm at the most, and SAF 2507 was slightly better. The most surprising observation was, however, that all parallel specimens of the old samples of both materials resisted the most severe conditions. This observation is attributed to an ageing effect of the oxide becoming more resistant at prolonged exposure in chlorinated seawater. The mechanism of this ageing is not studied here, but it is related to the effect to changes of the oxide thickness and composition caused by the repeated potential rise.

The consequence of this observation is that fresh samples should always be used when changing the testing conditions if one aims to obtain the lower limit of the acceptable environmental conditions. On the other hand, a practical consequence of the ageing effect is that the most critical phase for such materials in a chlorination plant is the first few weeks of

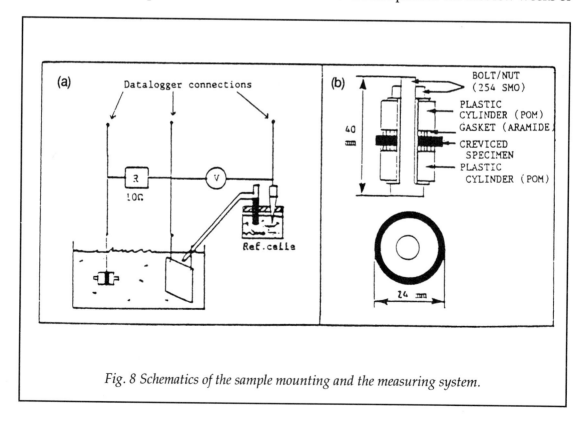

Fig. 8 Schematics of the sample mounting and the measuring system.

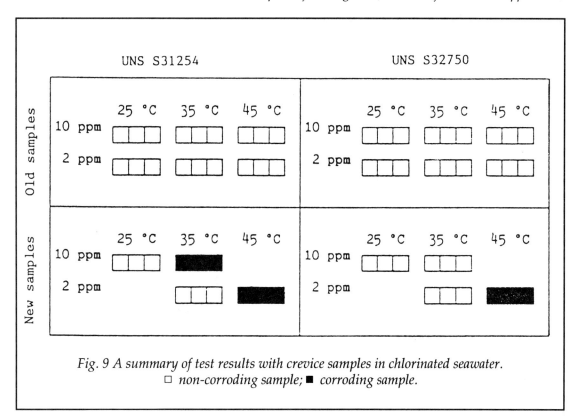

*Fig. 9 A summary of test results with crevice samples in chlorinated seawater.
☐ non-corroding sample; ■ corroding sample.*

operation. If crevice corrosion attack can be avoided early on, the chances of suffering corrosion attack at a later stage will be very small, provided that the environmental parameters and conditions do not become more severe.

Up to now we have referred only to the potential level when discussing the local corrosion susceptibility. In chlorinated seawater the potential rise takes place much faster than in natural seawater. As shown in Fig. 10(a) the potential rise in chlorinated seawater occurs within a few hours, as compared to a few days in natural seawater. This rapid potential rise has a direct influence on the local corrosion initiation tendency, at least for crevices. The reason is that the potential rise, dE/dt, adds an extra contribution to the passive current density [8]. With a larger passive current density there is an increased probability of crevice corrosion initiation. This is demonstrated in Fig. 10. The results of Fig. 10 are from an experiment in which three parallel specimens were exposed to a NaCl solution at 35°C with no chlorine present [8]. The potential rise with 2 ppm and the more rapid potential rise with 10 ppm chlorine was simulated using a potentiostat and a computer. As shown in Fig. 10(c), two out of three specimens subjected to fast potential scan initiated, while none of the slow scan specimens initiated. It should be noted that the final potential level was the same in both cases (600 mV SCE).

This experiment also has an important implication for corrosion testing. It shows that corrosion initiation has a maximum probability during the potential rise period. This is in agreement with other observations that initiation normally takes place within a rather short time in chlorinated seawater (1–2 d), if it ever occurs [24, 25]. For good reproducibility and reliability of the test results the experiment should be designed so that there is no unintended reduction of the test severity. This would be the case if the chlorine was added to cold water which was later heated to the test temperature over a time longer than the time of the potential rise, or, if the water was heated, but the chlorine concentration rose gradually over a period longer than the typical time of the potential rise. Such a smooth adjustment of the conditions

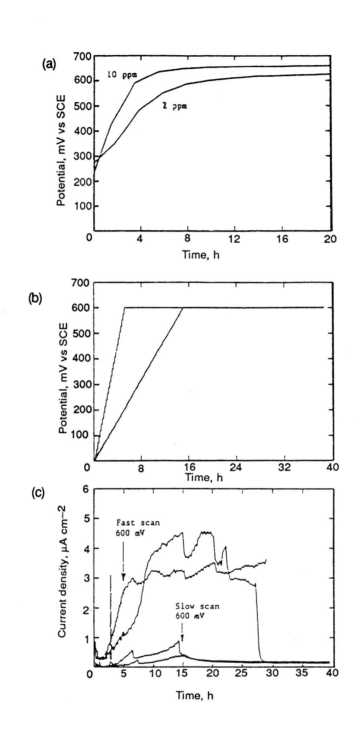

Fig. 10 (a) The development of the potential of non-corroding samples of 254 SMO in seawater at 35°C with 2 and 10 ppm chlorine [8]. (b) Fast and slow potential rise from 0 to 600 mV SCE simulated with a computer controlled potentiostat(c) Anodic current densities for two of three parallel crevice samples of 250 SMO exposed to a 3% NaCl-solution at 35°C. The three samples in each test series were subject to the two different potential developments as given in (b).

could be utilised as a positive remedy to reduce the risk of corrosion during the start-up phase in real systems carrying chlorinated seawater.

Turning now to the aspect of crevice corrosion propagation in chlorinated seawater, the cathodic efficiency is as important here as it was in natural seawater. Figure 11 shows some cathodic polarisation data obtained in flowing chlorinated seawater [12]. Here we observe that the influence of the chlorine is mainly limited to the very high potential region, where a relatively small current density is added to the oxygen reaction. In the propagation phase the potential will be lowered to somewhere in the region –200 to +200 mV SCE. In this region the cathodic efficiency in chlorinated seawater is much less than in natural seawater with an active biofilm (Fig. 3). This implies that, as long as the corrosion rate is controlled by the cathode, the corrosion rates will be lower in chlorinated seawater than in natural seawater up to 32°C. This is clearly verified by the results in Fig. 12, where an active crevice assembly of AISI 316 was coupled to a remote cathode with a fixed area ratio of 100:1 [25]. In the same way as shown previously in Fig. 5, the corrosion rate in natural seawater increases up to 1–3 mm/year, while it drops to the much lower level of about 0.1 mm/year at the end of the chlorination period. A qualitatively similar observation was made by Kain and Klein [26], comparing corrosion rates of alloy 625 in natural and chlorinated seawater at low area ratios.

In a real chlorination system, comprising extended lengths of pipes, tanks, heat exchangers etc., the area ratio between a corroding site and the cathode can be quite large. Simulation calculations carried out recently [11,12] have shown that in a large pipe system the effective area ratio in chlorinated seawater can be of the order 10 000:1, while in natural seawater maximum corrosion rates were achieved with an area ratio of less than 10:1. This difference in behaviour should be borne in mind when designing propagation tests and analysing propagation test results. With the relatively small area ratios used in testing (typically 100:1 or lower) corrosion rates measured in chlorinated seawater will tend to be lower than in a real system.

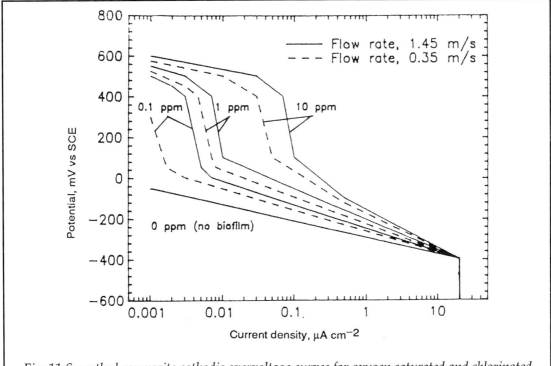

Fig. 11 Smoothed composite cathodic overvoltage curves for oxygen saturated and chlorinated seawater at 25°C.

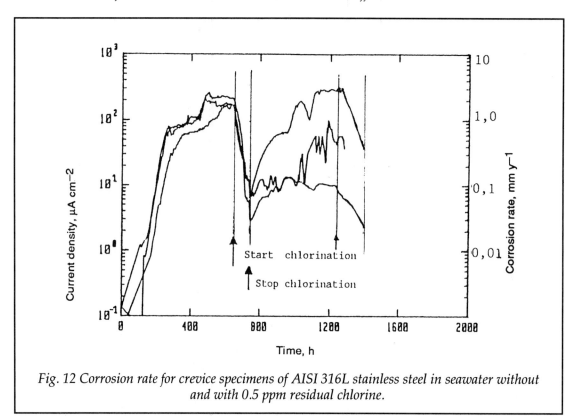

Fig. 12 Corrosion rate for crevice specimens of AISI 316L stainless steel in seawater without and with 0.5 ppm residual chlorine.

In natural seawater up to about 32°C the test results will be much more representative for a large plant.

1.3 Short term testing

Short term testing is needed as a supplement to exposure tests. The main reason for this is that exposure tests are difficult to standardise, particularly when conducted in natural seawater. Besides, they are quite expensive, time-consuming and cannot be run in most laboratories. Latterly, the critical temperature test has become quite popular for ranking materials. The most well known such test is the one based on testing in a $FeCl_3$ solution (ASTM G 48 and further developments of this test). The test can be used to define critical temperatures for pitting (CPT) as well as critical temperatures for crevice corrosion (CCT). The main advantage of this test is its simplicity, but there are serious drawbacks related to it, e.g. the definition of corrosion initiation. A different type of critical temperature test has been used a lot at SINTEF. Here, a potentiostat is used to control the potential at the desired level, while the specimen is exposed to a seawater-like electrolyte (3 % NaCl-solution). A typical experimental arrangement with a weld specimen is shown in Fig. 13. This method has several advantages compared to the ferric chloride test. The most outstanding are:

(i) The onset of corrosion can be well defined from measurements of the anodic current.
(ii) The potential level can be chosen to simulate the real environment (e.g. 400 mV SCE in natural seawater, 600 mV SCE in chlorinated seawater).
(iii) The electrolyte can be varied, if desired.

The major drawback in the use of a potentiostat is the possibility of erroneous behaviour of the electronics or the reference electrode. Besides, there is less standardisation associated with this test.

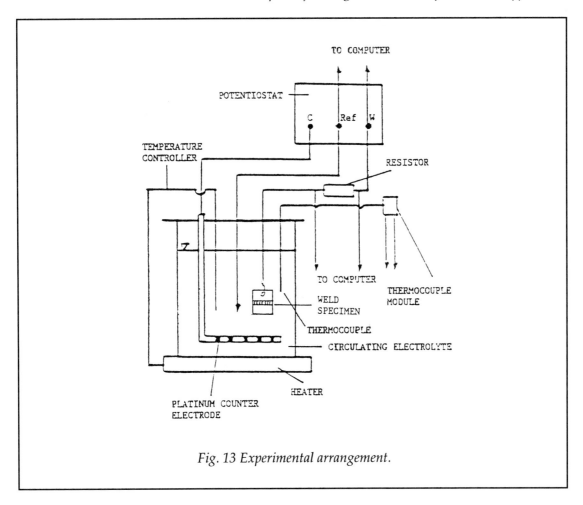

Fig. 13 Experimental arrangement.

A particularly interesting aspect of using the 'SINTEF test' is that a kind of 'design curve' can be obtained for a given material, where CPT (or CCT) is given as a function of the potential. This curve can then be used as a basis for recommendation of a given material in a given environment, as long as the expected maximum potential of the material in the environment can be defined. This requires, however, that the CPT (or CCT) values are 'true' values and not only relative values on a scale used for ranking purposes. Figure 14 shows examples of such 'design curves' obtained with the three materials AISI 316, SAF 2205 and 254 SMO. For the latter two materials there is one curve for pitting on the base material and one for pitting on a weld. The test method with crevice samples is still under modification, but it is expected that proper design curves with crevices will be close to the weld curves. Knowing the expected potential in the application environment (chlorinated seawater, natural seawater, formation water, etc.) the most critical design curve for a material (in Fig. 14, the weld curve) will indicate the maximum operating temperature the before onset of local corrosion. Such design curves cannot replace testing in the true environment, but can be used as an efficient screening test, thereby reducing the number of relevant materials.

2. Conclusion

1. For temperatures up to 32°C a biofilm stimulating the cathodic reaction is formed on stainless steel surfaces exposed to natural seawater. The accompanying potential rise to 300–400 mV SCE that takes place within a few days increases the risk of local corrosion initiation. The

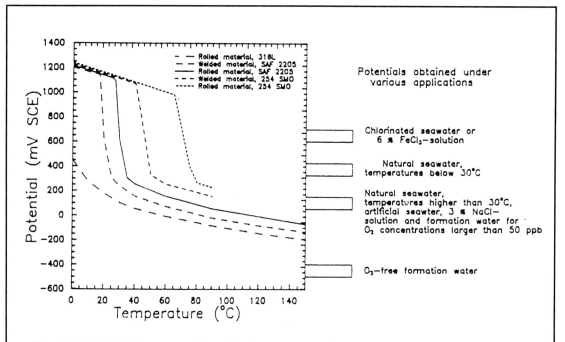

Fig. 14 Critical pitting potential–critical temperature diagram for base material and welded specimens of SAF 2205 and 254 SMO and for base material of AISI 316L. The free corrosion potential in various environments is also included.

enhancement of the cathodic efficiency results in anodically controlled propagation rates for very low area ratios.

2. The cathodic current density of stainless steel polarised to –100 to +100 mV SCE is a very sensitive indicator of the bio-activity on the surface.

3. In chlorinated seawater the potential of non-corroding stainless steel rises to the region 550–650 mV SCE within a few hours, thereby increasing the risk of local corrosion initiation further compared to natural seawater. The cathodic efficiency in chlorinated seawater is much less than in natural seawater up to 32°C. Propagation rates in chlorinated seawater will therefore be under cathodic control unless one is using very large cathode/anode area ratios.

4. The potential rise period is very important for the susceptibility to local corrosion initiation.

5. Upon prolonged exposure in chlorinated seawater the stainless steels become more resistant to local corrosion initiation.

6. A potentiostatically controlled critical temperature test is suggested to establish a design curve for maximum operating temperatures for a given stainless steel in seawater-like solutions.

3. Acknowledgements

This paper reviews work that has been carried out at SINTEF Corrosion Center over many years, and with contributions from many co-workers, in particular from Einar Bardal, John Drugli, Trond Rogne, Unni Steinsmo, Ragnar Holthe and Stein Valen. Financial support given by The Norwegian Council for Scientific and Technical Research (NTNF), The Norwegian Petroleum Directorate and the industry companies Statoil, Norsk Hydro, Stavanger Staal, Sandvik Steel and Avesta is gratefully acknowledged.

References

1. R. Johnsen, E. Bardal and J. M. Drugli, Cathodic Properties of Stainless Steel in Sea Water, Proc. 9th Scand. Corrosion Congr., Copenhagen, Sept. 1983.
2. R. Johnsen and E. Bardal, Cathodic Properties of Different Stainless Steels in Natural Seawater, Corrosion,1985, **41**, 5.
3. R. Holthe, E. Bardal and P. O. Gartland, Time dependence of cathodic properties of materials in seawater, Materials Performance, 1989, **28**, (6), 16.
4. R. Holte, P. O. Gartland and E. Bardal, Oxygen Reduction on Passive Metals and Alloys in Seawater—Effect of biofilm. 7th Int. Congr. on Marine Corrosion and Fouling, Valencia, Spain, Nov. 1988.
5. R. Holte, The Cathodic and Anodic Properties of Stainless Steels in Sea Water. Dr. ing. thesis, 1988, Dept. of Materials and Processes at the Norwegian Institute of Technology, University of Trondheim.
6. R. Gundersen, B. Johansen, P. O. Gartland, I. Vintermyr, R. Tunold and G. Hagen, The effect of sodium hypochlorite on bacterial activity and the electrochemical properties of stainless steels in seawater. Corrosion NACE '89, New Orleans, paper 108.
7. P. O. Gartland and J. M. Drugli, Crevice Corrosion of High Alloyed Stainless Steels in Chlorinated Seawater — Practical Aspects. Corrosion NACE '91, Cincinnati, paper 511.
8. P. O. Gartland and S. I. Valen, Crevice corrosion of high-alloyed stainless steels in chlorinated seawater — II, aspects of the mechanism. CorrosionNACE '91, Cincinnati, paper 510.
9. P. O. Gartland, R. Holte and E. Bardal, Evaluating crevice corrosion susceptibility from testing in simulated crevice electrolytes and mathematical modelling. 11th Scand. Corrosion Congr., Stavanger 1989.
10. S. I. Valen and P. O. Gartland, Critical temperatures for crevice corrosion of high alloyed stainless steels in sea water. Eurocorr '91, Budapest 1991.
11. S. I. Valen, Initiation, Propagation and Repassivation of Crevice Corrosion of High-Alloyed Stainless Steels in Seawater. Dr. ing. thesis 1991, Dept. of Materials and Processes at the Norwegian Institute of Technology, University of Trondheim.
12. P. O. Gartland and J. M. Drugli, Methods for Evaluation and Prevention of Local and Galvanic Corrosion in Chlorinated Seawater Piplines, Corrosion NACE '92.
13. T. Rogne, J. M. Drugli and R. Johnsen, Testing for initiation of the crevice corrosion of welded stainless steels in natural seawater, Materials Performance, September 1987.
14. T. Rogne, J. M. Drugli and R. Johnsen, Corrosion Testing of Welded Stainless Steel in Sea Water, CorrosionNACE '86, Houston, paper 230.
15. A. Mollica and A. Trevis, The Influence of Microbiological Film on Stainless Steels in Natural Sea Water, 4th Int. Congr. on Marine Corrosion and Fouling, Juan-les-Pins, 1976.
16. V. Scotto, R. Di Cinitio and G. Marcenaro, The influence of marine aerobic microbial film on stainless steel corrosion behaviour, Corros. Sci., 1991, **25**, 3.
17. J. P. Adudonard, A. Desestret, L. Lemoine and Y. Morizur, Special Stainless Steels for Use in Seawater, UK Corrosion, Wembley, 1984.
18. S. C. Dexter and G. Y. Gao, Effect of Seawater Biofilms on Corrosion Potential and Oxygen Reduction of Stainless Steels, Corrosion NACE '87, San Francisco, paper 377.
19. P. Gallager, R. E. Malpas and E. B. Shone, Corrosion of stainless steels in natural, transported and artificial seawaters, Brit. Corros. J., 1988, **23**, 4.
20. A. Mollica, A. Trevis, E. Traverso, G. Ventura, G. de Carolis and R. Dellepiane, Cathodic Performance of Stainless Steels in Natural Seawater as a Function of Microorganism Settlement and Temperature, Corrosion, 1989, **45**, 1.
21. V. Scotto, G. Alabiso and G. Marcenaro, Bioelectrochemistry and Bioenergetics (1986) BEBC 0945, 1986.

22. F. Mansfeld and B. Little, A technical review of electrochemical techniques applied to microbiologically influenced corrosion, Corros. Sci., 1991, **32**, (3), 247.
23. R. Johnsen, Corrosion failures in sea water piping system offshore, this volume, pp.48–58.
24. B. Wallen and S. Henrikson, Effect of Chlorination on Stainless Steels in Seawater, Corrosion NACE '88, St. Louis, 1988, paper 403.
25. T. Rogne, to be presented at NACE Conference "Engineering Solutions to Materials Problems in the Oil Industry", Sandefjord, Norway, 7–9 June, 1993.
26. R. M. Kane and P. A. Klein, Crevice corrosion propagation studies for alloy N06625: Remote crevice assembly testing in flowing natural and chlorinated seawater, CorrosionNACE '90, Las Vegas, paper 158.

13

Biofilm Monitoring in Seawater

A. Mollica, E. Traverso and G. Ventura

Istituto per la Corrosione Marina dei Metalli (ICMM-CNR), Via De Marini 6/8, 16149 Genova, Italy

Abstract

The phenomenon of oxygen reduction depolarisation induced by biofilm growth on a series of active–passive alloys (stainless steels, Ni–Cu, Ti, Ni–Cr ....) while exposed to seawater was exploited for developing biofilm monitoring devices to be applied, in particular, in condensers.

The results of field tests show that the measure of the galvanic current, in a very simple device consisting of a stainless steel pipe coupled with a sacrificial iron anode, is a promising method for the monitoring of thin biofilm layers (of less than 50µm) on pipe walls, and therefore in a range of biofilm thickness not detectable by the friction factor or thermal exchange resistance measurements.

Such a device is able to show clearly, in real time, the effect of chemical and mechanical cleaning procedures on thin biofilm layers, and can, therefore, be used to monitor and optimise antifouling procedures.

Practical implications of electrochemical methods for both biofilm monitoring and microbially induced corrosion prevention are also discussed.

1. Introduction

Biofilm growth on condenser pipe walls is the most important cause of efficiency loss in power plants[1, 2]. The additional cost due to the presence of fouling in condensers, has been estimated at several billions of dollars per year in the USA alone.

The fight against the detrimental effects of the biofilm is, therefore, imperative and for this purpose antifouling procedures, mainly based on the addition to seawater of chemicals, like chlorine, are widely applied; on the other hand the use of chemicals tends to be reduced by law, because of their ecological impact.

"This situation poses a real dilemma. Utilities are expected to refine their control procedures when little is known about the fundamental mechanisms of biofouling or its control, and when no means are available for predicting or accurately detecting the onset of condenser biofouling. Utility operators and designers urgently need....reliable methods or devices for rapid, on-line detection of biological activity and growth....[3]".

The aim of the devices is, obviously, to give information which can be utilised for optimising the antifouling control procedures.

A number of devices for biofilm monitoring are indeed on the market [4] and others have been recently proposed [5–7] particularly those which are easy to handle and able to give, on line and in real time, information about the biofilm development. Furthermore, they provide an indirect evaluation of the biofilm thickness based on accurate measurements of the thermal exchange and/or the frictional resistance.

For a number of reasons, these kinds of devices are not able to indicate clearly the presence of a biofilm layer thinner than about 50μm[8–9]. It follows that, if these devices are used to monitor antifouling control procedures, a surface still settled by bacteria will be considered as 'clean': the residual microbial population, remaining from previous treatments, will rapidly regrow and the biofilm resistance to further antifouling procedures will increase[10–12].

Methods for detecting very thin biofilm layers are therefore useful and, for this purpose, some tests were carried out at the ICMM to develop a biofilm monitoring device based on the following effects of the biofilm growth.

Thus, it has been observed that the growth of a biofilm on a variety of active–passive alloys (stainless steels, Ni–Cr, Ni–Cu, Ti,....) induces a depolarisation of the oxygen reduction on the metallic surfaces, and the effects of this phenomenon on the corrosion of these alloys and less noble materials coupled to them has been studied[13–15].

For the purpose of biofilm monitoring, this effect can be rephrased as: the change in time of the oxygen reduction kinetics can be used as an index of the biofilm growth.

Looking at the scheme in Fig. 1, information on the oxygen reduction evolution, linked to the biofilm growth, can be easily obtained in, at least, two different ways; viz., by following the potential change at a fixed cathodic current (i_c), and by following the cathodic current change at a fixed potential (E_c).

For practical applications, these two techniques (intentiostatic and potentiostatic) can be approximated by following the free corrosion potential development of the alloy in the passive state ($i_c = i_p$), and by following, with time, the galvanic current (I_g) flowing between an active–passive alloy and a suitable sacrificial anode.

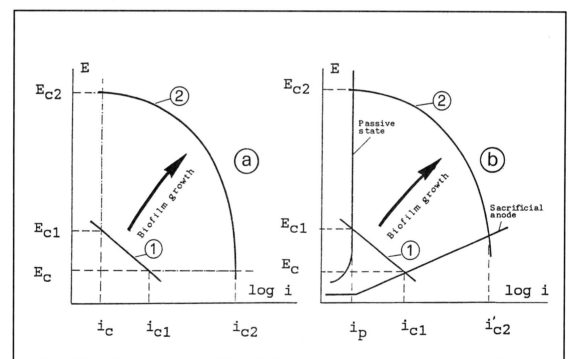

Fig. 1 Schematic representation of the cathodic curves showing the oxygen reduction kinetics on active–passive alloys; (1) in absence of biofilm, and (2) in presence of biofilm on their surfaces, and possible exploitation of this phenomenon in order to obtain index of the biofilm growth:
(a) by intentiostatic or potentiostatic technique, and (b) in practical approximation of these techniques.

The possible use of this last method as a biofilm monitoring system will be considered in this paper.

In particular, the aim of this work is to ascertain if the very simple measurement of the galvanic current between a stainless steel pipe and a sacrificial iron anode is able to achieve the following:

(i) signal clearly the presence on the pipe wall of a biomass amount not detectable by friction factor measurements;

(ii) follow in real time the effect of antifouling procedures such as continuous or intermittent chlorine additions and mechanical cleaning procedures;

(iii) give, at the same time, information which can be used to prevent the risk of microbial induced corrosion (MIC).

2. Experimental

Possible biofilm monitoring devices, loops (as shown in Fig. 2), were studied in which natural seawater is pumped once through stainless steel pipes coupled together to simulate a single pipe and, together, with an sacrificial iron anode.

By this arrangement, the potential at the mouth and at definite points inside the pipe can be measured as well as the galvanic current between iron and pipe.

The seawater, before entering the pipe, can be treated by continuous or intermittent NaClO additions to control the biofilm growth or remove a pre-existent biofilm on the pipe wall.

During the tests, friction factor measurements and protein analysis were also made on additional pipe elements to correlate the electrochemical signals with different biofilm monitoring systems.

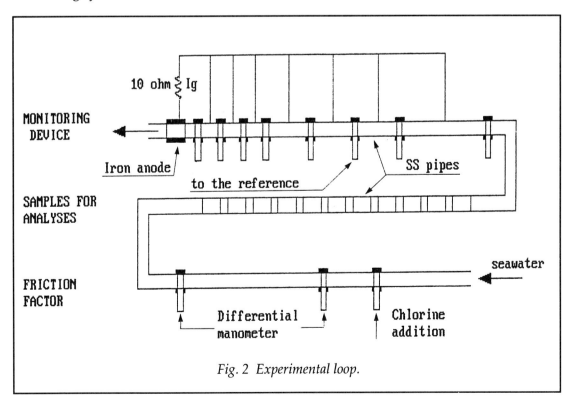

Fig. 2 Experimental loop.

3. Results

3.1 Continuous NaClO additions

In a first test, lasting about 35 days, 6 loops were used to monitor the biofilm growth in presence of different continuous NaClO additions; the mean continuous residual chlorine concentrations in the tested loops were 0, 0.05, 0.1, 0.2, 0.4 and 0.8 ppm respectively [7].

The graphs in Fig. 3(a)–(d) summarise the results that were obtained and support the following conclusions.

At the end of the test:

(i) the friction factor measurements (Fig. 3(a)) were able to indicate, clearly, the biofilm presence only in untreated seawater: this means that, in the presence of continuous chlorine additions, the biofilm on the pipe walls had reached a thickness certainly lower than about 50μm;
(ii) the measurements of the protein amount (Fig. 3(b)) were more sensitive and able to reveal the existence of a biomass also in the presence of 0.05 ppm of residual chlorine;
(iii) no significant biomass amount was detected for the higher tested chlorine concentrations, in agreement with the results of other authors [46–48].

During the test, the measurements of the galvanic currents (Fig. 3(c)) indicate:

(i) a sharp increase of the galvanic current during the first 7–10 days, followed by a plateau, in untreated seawater indicating that the oxygen reduction depolarisation on stainless steel pipes, causing the increase of the sacrificial anode consumption, is mainly linked to the first phase of the biofilm growth;
(ii) a lowering of the rate of current increase in presence of 0.05 ppm of continuous residual chlorine;
(iii) no change in time for residual chlorine concentration over 0.1 ppm.

The galvanic currents measured at the end of the exposure are plotted vs the residual chlorine concentration (Fig. 3(d)) and the resulting graph can be compared with (Fig. 3(a, b)), to examine the possibility of using this electrochemical signal for the monitoring of the biofilm growth.

The following conclusions can be drawn:

(i) an increase of one order of magnitude of the electrical signal is observed for the small amount of biomass grown in presence of 0.05 ppm residual chlorine and, therefore, long before the biofilm layer reaches a thickness detectable by friction factor measurements (about 50μm): the signal generated by such a biofilm layer is, therefore, of wide amplitude and easy to measure;
(ii) the signal is almost insensitive to a further increase of the biofilm thickness;
(iii) the galvanic current measurements could be used as a very early warning of the biofilm growth: in fact at 0.1 ppm of residual chlorine, the electrical signal appeared to be able to reveal the presence of a biomass amount that was not clearly detected by protein measurements;
(iv) when no biofilm is present on the pipe wall, the signal remains effectively stable during the test at a value almost independent of the residual chlorine concentration that is necessary to avoid the biofilm adhesion (at least for residual chlorine concentrations between 0.2 and 0.8 ppm).

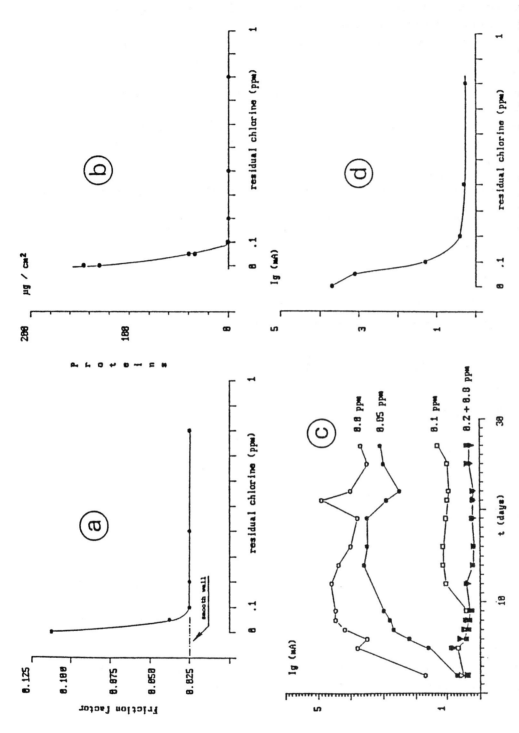

Fig. 3 Results of a test performed in 6 loops fed with seawater treated by different and continuous chlorination levels in the range 0 to 0.8 ppm of residual chlorine [7]. (a) Friction factor vs chlorine at the end of the test; (b) Protein amount on the pipe walls at the end of the test; (c) Trends of the stainless steel-iron galvanic current during the test for different chlorination levels; (d) Final values of the galvanic currents versus residual chlorine concentration.

3.1.1 Summary

The proposed device appears able to signal clearly, in real time and without interference from the chlorine concentration, the presence of thin biofilm layers in a range of thickness not revealed by friction factor measurements.

Additional information obtained in these tests points to a link between the presence of a biofilm and the protection of stainless steel pipes by sacrificial anodes.

In Fig. 4 the potential profiles measured at the end of the test inside the stainless steel pipes protected by an iron anode and exposed to seawater treated with different chlorine amounts are plotted.

It can be seen that the high protection current required by the fouled pipe exposed to untreated seawater causes a high ohmic drop at the inlet of the pipe and consequently, only a short pipe length, of the order of some 10 cm, will be protected.

The elimination of the biofilm by a suitable chlorine concentration not only reduces the sacrificial anode consumption but, in addition, causes the levelling out of the potential inside

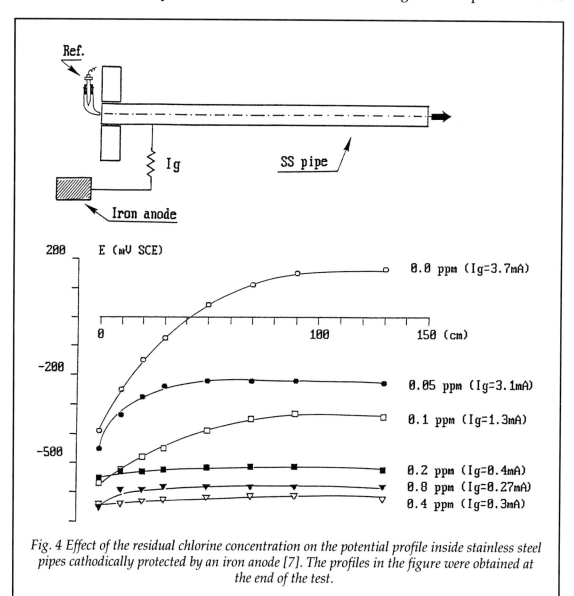

Fig. 4 Effect of the residual chlorine concentration on the potential profile inside stainless steel pipes cathodically protected by an iron anode [7]. The profiles in the figure were obtained at the end of the test.

the pipe to a very protective value and, therefore, the protected pipe length increases to a value of several meters.

In other words, measurements of galvanic currents provide information not only about the presence of thin biofilm layers on the pipe walls but, at the same time, also about the pipe length protected against MIC.

3.2 Intermittent NaClO additions

A second test, lasting about 80 days, was conducted to check if the measurements of galvanic current could also be used to indicate the elimination of a pre-existent biofilm when intermittent NaClO additions are employed.

The graph in Fig. 5(a) shows the chlorination program applied to six loops similar to that previously described; the graph in Fig. 5(b) shows the trend of the stainless steel pipe-iron galvanic currents measured during the test in each loop.

The comparison of the two graphs indicates that some reduction of the electrical signal can be observed when chlorine, at whatever concentration, is added to seawater but, if it is necessary to detach rapidly a biofilm already present on the stainless steel pipe walls, a treatment with a residual chlorine concentration over 0.5–0.6 ppm will be necessary. This conclusion is in agreement with the findings of other authors[46, 49].

Finally, an additional test was performed using the monitoring device in which the measurements of the galvanic current were directly expressed in terms of biofilm mean thickness assuming, provisionally, a full scale of about 20μm.

The graph in Fig. 6 shows the result of this test: it can be seen how this device appears to be able to follow in real time the first phase of the biofilm growth, the effect on the growth rate of the change of some environmental parameters and, finally, the biofilm elimination by a shock chlorine addition.

In particular, it suggests that a residual chlorine concentration of 1.5 ppm is able to eliminate about 80% of the biofilm in 4 h; a very similar result was obtained by Miller and Bott [50].

3.3 Mechanical cleaning

A final test, lasting about 90 days, was performed to check if the measurements of the galvanic current could be applied to follow the effect of mechanical cleaning on a pre-existent biofilm [13].

In Fig. 7 the galvanic currents between an iron anode and a stainless steel pipe are plotted vs time. Here the biofilm was repeatedly removed by a nylon brush, with and without subsequent flushing with ethyl alcohol.

The trend of the galvanic current indicates that the mechanical cleaning alone is not able completely to clean the pipe wall (the ordinates of the points C and E are higher than the ordinate of the initial point A) so that the regrowth is very fast; a clean surface was restored only when the mechanical cleaning was followed by a chemical cleaning (compare the ordinates of the points A, G, I).

Similar results were obtained by different authors [10–12].

4. Discussion

The phenomenon of the oxygen reduction depolarisation induced by biofilm growth on a series of active–passive alloys exposed to seawater was exploited for the development of biofilm monitoring devices.

In particular, the results of our tests show that the measure of the galvanic current in a very simple device consisting of a stainless steel pipe coupled to a sacrificial iron anode is a promising method for the monitoring of thin biofilm layers on the pipe walls in a range of

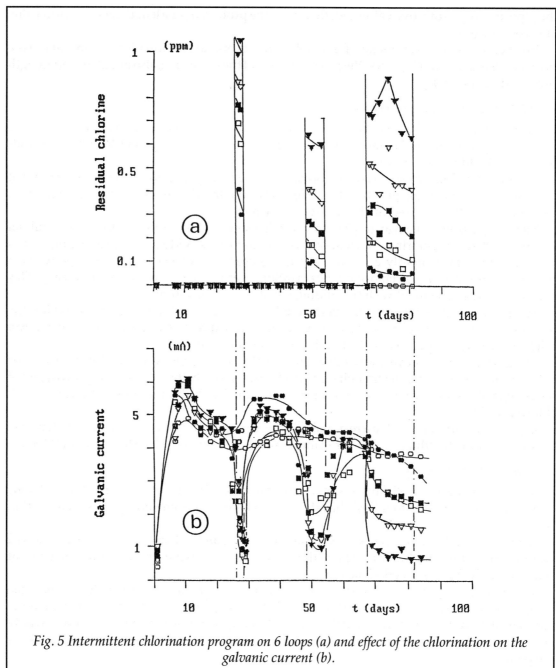

Fig. 5 Intermittent chlorination program on 6 loops (a) and effect of the chlorination on the galvanic current (b).

thicknesses not detectable by friction factor or thermal exchange resistance measurements.

The information obtained by this simple device is, in fact, in agreement with the results obtained in different ways by other authors in relation to the effect of mechanical and chemical antifouling procedures; in addition, such a device is easy to handle, suitable for on line and real time biofilm monitoring, and able to signal biofilm growth up to a thicknesses certainly lower than 50 μm by a signal of wide amplitude (increase of about one order of magnitude of the galvanic current).

For practical application in a condenser, we should evaluate whether the condenser may be considered, by itself, as a biofilm monitoring device on the tube nest walls.

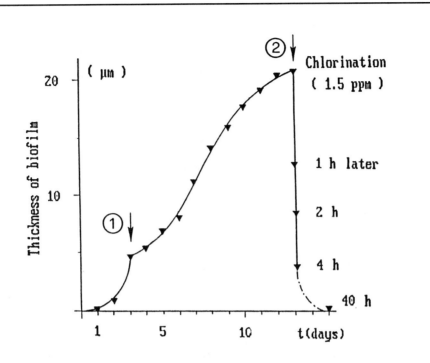

Fig. 6 Biofilm growth and removal as suggested by the monitoring device. The arrow (1) indicates when rain caused a small temporary decrease of the salinity and temperature of the feed seawater; the arrow (2) shows when chlorine, at a level of 1.5 ppm of residual chlorine, was added in seawater.

In effect, cathodic protection systems, by sacrificial anodes or impressed currents, are very often applied in the water box and, therefore, signals equivalent to the measurement of the galvanic current utilised in our device can be directly obtained *in situ*.

Some indirect positive answers regarding this possibility can be found in the literature. Muller, Lang and Healy [51] observed the mixed potential inside the water box of a condenser protected by sacrificial iron anodes and with the tube nest made of stainless steel.

They noted that regular monitoring of the mixed water box potential revealed that a decrease in the level of protection occurred within a few weeks, long before the sacrificial anodes were consumed. Surprisingly, the good function of the iron anodes was restored after an intermittent chlorination. This remarkable behaviour could be repeated many times. Furthermore, the observed ennoblement of the mixed potential during the exposure to untreated seawater, which was then erased by the chlorination, was close to 150 mV; a very similar shift can be observed in our device (Fig. 4) when comparing the potential at the mouth of fouled and unfouled stainless steel pipes.

It seems therefore that the measure of the mixed potential can be used directly in a condenser, as a substitute for the galvanic current procedure for the monitoring of the biofilm growth and removal.

A second observation was made by Nekoksa and Hanson[52] who followed the behaviour of a protection system by impressed current applied to a condenser with titanium pipes and stainless steel pumps. They observed a gradual increase (up to 5 times) of the cathodic protection current in several weeks after a decrease or cessation of the chlorine addition.

These data are essentially in agreement with the results of our monitoring device (Fig. 3 and 5) confirming that electrochemical methods can be directly applied *in situ* for biofilm monitor-

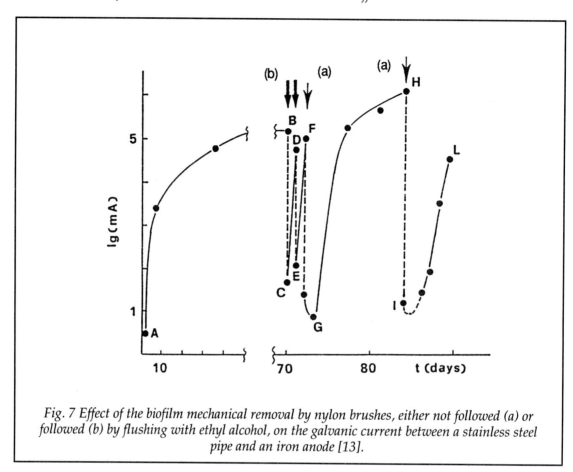

Fig. 7 Effect of the biofilm mechanical removal by nylon brushes, either not followed (a) or followed (b) by flushing with ethyl alcohol, on the galvanic current between a stainless steel pipe and an iron anode [13].

ing on tube nest walls not only made of stainless steel, but in general, of the other active–passive alloys mentioned above.

5. Conclusions

1. Information about the presence of a thin biofilm layer on active–passive alloys (stainless steels, Ni–Cu, Ti, Ni–Cr) exposed to seawater can be easily obtained by following the trend of the galvanic current flowing between these alloys and a suitable sacrificial anode: the change in time of the galvanic current is one of the consequences of the phenomenon of the oxygen reduction depolarisation induced by biofilm growth.

2. A simple biofilm monitoring device based on galvanic current measurements which uses an iron anode as a sacrificial element is able to provide information on the first phases of biofilm growth up to a maximum thickness of about 30-40 µm.

3. The growth of the biofilm in this range is signalled by an increase of one order of magnitude of the electrical signal.

4. This device is able to show, in real time and on line, the effect of chemical and mechanical cleaning procedures on thin biofilm layers, and can, therefore, be utilised to monitor and optimise antifouling procedures.

5. In a condenser, the removal of the biofilm by a well controlled antifouling procedure, causes, at the same time, a reduction of the sacrificial anode consumption and a deeper throw of the cathodic protection inside the pipe.

References

1. P. J. Battaglia et al., Biofouling Control Practice Assessment, EPRI, CS1796, April 1981.
2. J. Graham, Biofouling Control Assessment — A Preliminary Data Base Assessment, EPRI, CS-2469, July 1982.
3. W. Chow, Overview, Proc. Condenser Biofouling Control Symp.: the State of the Art, EPRI, Florida, June 1985.
4. D. Anson, J. Corliss, B. Vigon and R. Hillman, Biofouling Monitors for Condenser Tube Applications, Proc. Condenser Biofouling Control Symp.: the State of the Art, EPRI, Florida, June 1985.
5. D. E. Nivens, J. Q. Chambers and D. C. White, Non-Destructive Monitoring of Microbial Biofilms at Solid-Liquid Interface Using On-Line Devices, Proc. Microbially Influenced Corrosion and Biodeterioration, Knoxville, Tennessee, October 1990.
6. G. H. Markx and D. B. Kell, Biofouling, 1990, **2**, 211.
7. A. Mollica, E. Traverso and G. Ventura, Proc. 11th Int. Corrosion Congr., Florence, Italy, Vol. 4, p. 341, April 1990.
8. R. Sugam, W. A. Sandvik and H. S. Arnold, Design and Operation of a Condenser Fouling Monitor, Proc. Condenser Biofouling Control Symposium: the State of the Art, EPRI, Florida, June 1985.
9. W. G. Characklis, Biofilm Development and Destruction, EPRI, CS-1541, Research Project 902-1, September 1980.
10. R. O. Lewis, Corrosion, 1982, **38**, (9), 31.
11. J. S. Nickels, I. H. Parker, R. J. Bobbie, R. F. Martz, D. F. Lott, P. H. Benson and D. C. White, Int. Biodet. Bulletin ISSM 0020-616417(3) Autumn 1981.
12. R. J. Bobbie, D. C. White and P. H. Benson, Proc. 5th Int. Congr. on Marine Corros. and Fouling, Marine Biology Barcelona, May 1980.
13. A. Mollica, A. Trevis, E. Traverso, G. Ventura, V. Scotto, G. Alabiso, G. Marcenaro, U. Montini, G. Decarolis and R. Dellepiane, Proc. 6th Int. Congr. on Marine Corrosion and Fouling, Athens, Greece, p.269, September 1984.
14. A. Mollica and A. Trevis, Proc. 4th Int. Congr. on Marine Corros. and Fouling, Antibes, France, 1976, 351.
15. J. M. Krougman and F. P. IJsseling, Proc. 5th Int. Congr. on Marine Corrosion and Fouling, Barcelona, Spain, 1980, 214.
16. E. D. Mor, V. Scotto and A. Mollica, Werkstoffe u. Korros., 1980, **31**, 281.
17. T. S. Lee, R. M. Kain and J. W. Oldfield, Materials Perfomance, 1984, **23**, (7), 9.
18. D. J. Schiffrin and R. S. De Sanchez, Corrosion, 1985, **41**, (1), 31.
19. V. Scotto, R. Di Cintio and G. Marcenaro, Corros. Sci., 1985, **25**, (3), 185.
20. R. Johnsen and E. Bardal, Corrosion, 1985, **41**, (5), 296.
21. H. Arup, Electrochemical Techniques in Marine Corrosion Research, Proc. 8th European Congr. of Corrosion, Nice, CP7-1-11, 1985.
22. E. B. Shone and P. Gallagher, Proc. High Alloy Stainless Steels for Critical Seawater Application, Birmingham, UK, 1985, 27.
23. D. E. Nivens, P. D. Nichols, J. M. Henson, G. G. Geesey and D. C. White, Corrosion, 1986, **42**, (4), 204.
24. V. Scotto, G. Alabiso and G. Marcenaro, Bioelectrochemistry and Bioenergetics, 1986, **16**, 347.

25. B. Wallen and T. Anderson, Proc. 10th Scand. Corrosion Congr., Stockholm, Sweden, 1986, 149.
26. A. Desestret, Materiaux et Techniques, 1986, 317.
27. R. Johnsen and E. Bardal, Corrosion NACE '86, Houston, Texas, paper 227.
28. A. Mollica, G. Ventura, E. Traverso and V. Scotto, Int. Biodeterioration, 1988, **24**, 221.
29. B. Wallen and S. Henrikson, CorrosionNACE '88, St. Louis, Missouri, paper 403.
30. S. C. Dexter and G. Y. Gao, Corrosion, 1988, **44**, (10), 717.
31. P. Gallagher, R. E. Malpas and E. B. Shone, Brit. Corros. J., 1988, **23**, (4), 229.
32. A. Mollica, A. Trevis, E. Traverso, G. Ventura, G. Decarolis and R. Dellepiane, Corrosion,1988, **44**, (4), 194.
33. R. Holthe, P. O. Gartland and E. Bardal, Proc. 7th Int. Congress on Marine Corrosion and Fouling, Valencia, Spain, 1988.
34. G. Ventura, E. Traverso and A. Mollica, Proc. 1st European Federation Corrosion Workshop on Microbial Corrosion, Sintra, Portugal, 1988. Published by Elsevier, London.
35. R. Holthe, E. Bardal and P. O. Gartland, Corrosion NACE '88, St. Louis, Missouri, paper 393.
36. O. Varjonen, T. Hakkarainen, E. Nurmiaho-Lassila and M. Salkinoja-Salonen, Proc. 1st Workshop on Microbial Corrosion, Sintra, Portugal, 1988. Published by Elsevier, London.
37. A. Mollica, A. Trevis, E. Traverso, G. Ventura, G. Decarolis and R. Dellepiane, Corrosion, 1989, (1), 48.
38. G. Ventura, E. Traverso and A. Mollica, Corrosion, 1989, **45**, (4), 319.
39. S. Valen et al., Proc. 11th Scand. Corrosion Congr., Stavanger, Norway, Paper N F-36, 1989.
40. R. Holthe, E. Bardal and P. O. Gartland, Materials Performance, 1989, **28**, (6), 16.
41. B. Wallen and S. Henrikson, Werkstoffe und Korros., 1989, **40**, 602.
42. A. Mollica, G. Ventura and E. Traverso, Proc. Int. Congr. on Microbially Influenced Corrosion, Knoxville, Tennessee, USA, 1990.
43. V. Scotto, G. Alabiso, M. Beggiato, G. Marcenaro and J. Guezennec, Proc. 5th European Congr. on Biotechnology, Copenhagen, 1990.
44. A. Mollica, Materiaux Et Techniques, Special Biocorrosion, 1990, 17.
45. S. C. Dexter and H. J. Zhang, Proc. 11th Int. Corrosion Congr., Florence, Italy, 1990, 4, 333.
46. R. J. Soracco, E. W. Wilde, L. A. Mayack and D. H. Pope, Water Res., 1985, **19**, (6), 763-766.
47. P. D. Goodman, Brit. Corros. J., 1987, **22**, (1).
48. R. Gundersen et al., Proc. UK Corrosion with Eurocorr '88, Brighton, UK, 125.
49. R. J. Boley, Materials Performance, 1980, 31.
50. P. C. Miller and T. R. Bott, Progress in the Prevention of Fouling in Industrial Plant, Nottingham, UK, 1981, 121.
51. R. O. Muller, V. H. Lang and T. B. Healy, Monitoring of Biofouling and Cathodic Corrosion Protection, Condenser Biofouling Control Symposium: the State of the Art, EPRI, 1985.
52. G. Nekoksa and R. T. Hanson, Effects of Chlorination on Cathodic Protection Current Requirements, Corrosion NACE '89, paper 286.

MECHANISM

Identification of Sulphated Green Rust 2 Compound Produced as a Result of Microbially Induced Corrosion of Steel Sheet Piles in a Harbour

J.-M. R. Génin, A. A. Olowe, B. Resiak*, N. D. Benbouzid-Rollet[†], M. Confente* AND D. Prieur[†]

Laboratoire CNRS Maurice Letort UP 6851, Université de Nancy 1, Département Sciences des Matériaux, ESSTIN, Parc R. Bentz, F 54 500, Vandoeuvre-Nancy, France
*Unimétal Recherche, BP 140, F 54 360, Amnéville, France
[†]Station Biologique CNRS UP 4601, Place Teissier, BP 74, Roscoff-Cedex, 29 682, France

Abstract

Mössbauer spectroscopy and X-ray diffraction were used to show that the corrosion of steel sheet piles at the level of lowest tides in a harbour in the presence of sulphate reducing bacteria in anaerobic conditions, gives rise to green rust 2(GR2), $4Fe(OH)_2$, $2FeOOH$, $FeSO_4$, nH_2O, a ferrous-ferric sulphated compound, mixed with magnetite.

1. Introduction

The presence of green rust 2 (GR2) obtained by microbially induced corrosion of steel coupons due to sulphate reducing bacteria (SRB) in marine sediments has been recently reported [1]. In this short article, we shall confirm that from preliminary test results on the corrosion of steel sheet piles in the French harbour of Boulogne, the major product of corrosion (providing anaerobic conditions are maintained) is the sulphated GR2.

2. Experimental and Results

A number of European harbours suffer drastic corrosion of steel sheet piles. In the major facility at the harbour of Boulogne, France, the piles are 6 m high and constitute the foundations of the quay. The 1 cm thick steel sheets have been perforated at a height of 1.50 m from the natural bed of the harbour, i.e. at about 0.50 m above the level of the lowest tides. The material is plain carbon steel made by the Thomas process in the early 1960s. Even though perforations are always at the same level, it is qualitatively obvious that the corrosion is also very active all over the steel sheets several metres above the location of the perforations. The sheets are covered with a hard layer of shells, the thickness of which is very regular, i.e. of the order of 5 cm. This layer must provide an effective protection between the seawater and the steel surface and will ensure anaerobic conditions. Samples from 20 to 100 g in mass are scraped off from this layer in order to select corrosion products as close as possible to the remaining material. Two types of samples are taken: some for bacterial analysis and others for rust analysis by Mössbauer spectroscopy and X-ray diffraction. These last samples are immediately frozen into liquid nitrogen for preservation in order to avoid any spurious oxidation before the analysis is carried out.

2.1 Bacterial analysis

The samples of corrosion products are taken at different heights (two samples per height) above the sediment on the harbour bed which is taken as the reference level (0 m). They are immediately placed in bottles which contain 270 ml of natural seawater sterilised by autoclaving under nitrogen gas for 20 min at 120 °C and previously reduced by the addition of 1 g l^{-1} ascorbic acid in the presence of reazurine which is used as a deoxygenation indicator.

The bottles are maintained at 5 °C until the analysis is carried out in anaerobic conditions under an atmosphere of 90 % nitrogen, 5 % hydrogen and 5 % carbon dioxide. Numerical records are made by the method of the most probable number in a liquid medium which contains acetate and lactate as a carbon source, which is a formulation used for the enumeration of all the sulphate reducing bacteria [2]. The liquid medium is inoculated in triplicate with the sample water, after prolonged shaking of the bottles containing the samples. With the pH of the medium at 7.4 and a temperature of incubation of 20–22°C, values are adjusted to correspond to natural conditions as prescribed by Bak and Pfennig [3]. After one month of incubation, all samples are observed to contain sulphate reducing bacteria.

The presence of these anaerobic micro-organisms in the surrounding water even though oxygenated, can be attributed to a contamination by the underlying sediment, which is probably highly anoxic. Because of the dilution of the water samples, the bacteria concentrations at the contact with the metal are around 10^6 to 10^7 per gram of corrosion product. These high concentrations imply that conditions are very favourable for the multiplication of the sulphate reducing bacteria, i.e. anaerobic, with the presence of electron donors and acceptors and the absence of interspecific competition. The most dense population is situated at a height of 1.50 m from the sediments, which corresponds to the level of the perforation (Table 1).

2.2 The rust products

When the sample is scraped off the sheet pile the bright steel surface quickly becomes greenish after flushing the surrounding concretion layers. The samples are immediately preserved at 78 K for transportation before being studied by Mössbauer spectroscopy and X-ray diffraction. These analyses are also carried out in anaerobic conditions. Mössbauer spectroscopy is a unique method of investigation, since it is only sensitive to compounds which contain iron, and so is a very relevant method because the products which are analysed are taken from the black concretions on the vertical sheet piles between the steel surface and the external layer made of sea shells. Therefore, the low absorption percentage of the Mössbauer spectra, often smaller than 1%, is essentially due to the gangue which is mixed with the rust samples. Thus, some spectra are measured for as long as one week. However, all the spectra display the characteristic peaks found in spectra of the standard green rust 2 (GR2). This compound which was observed first in 1935 by Girard [4] has been thoroughly studied by A. A. Olowe [5–7]. The role of the sulphate ions in the oxidation of $Fe(OH)_2$ is described in this volume [8].

The formula of standard GR2 is established to be 4 $Fe(OH)_2$, 2 $FeOOH$, $FeSO_4$, nH_2O and the

Table 1 Numbers of sulphate-reducing bacteria at the contact of the metal expressed as the number of cells per ml of sampled water, for various heights above the sediment

Origin	Number of Bacteria
Surrounding water	1.1×10^3
height: 1.00 m	2.5×10^5
height: 1.50 m	6.0×10^5
height: 2.00 m	1.2×10^4
height: 3.00 m	3.2×10^5

corresponding Mössbauer spectra measured at 295 and 78 K are shown in Fig. 1(a) and (b). Five crystallographic sites are distinguished, three ferrous sites corresponding to three doublets D_1, D_2, D_3 and two ferric sites corresponding to two doublets D_4, D_5. They are in the intensity ratio $D_1/D_2/D_3/D_4/D_5 = 2/2/1/1/1$. The resolution at 295 K does not allow D_1 to be distinguished from D_2. However the Fe^{2+}/Fe^{3+} ratio is always 2/5 and GR2 is a well defined compound.

Figure 1(c) shows a spectrum of a sample taken at the height of 3 m (1.50 m above the perforation). The spectrum is measured at room temperature at a velocity scale of ± 11 mm s^{-1}, to be able to determine all iron components present. Note that the central paramagnetic part corresponds only to a maximum absorption percent of 1.5 %, the counting reaching 4.3×10^6. The sample comprises about 37 % of magnetite characterised by two sextets at 486 and 450 kOe, 5% of goethite and GR2 (Table 2). Only D_3 is distinguished from (D_1, D_2). The abundance of the ferric doublet shows that it is made of two ferric doublets (D_4, D_5) and some extraneous ferric components due to some rapid partial oxidation of GR2 [1, 9]. It can be deduced that the part allocated to GR2, i.e. $D_4 + D_5$ in the paramagnetic ferric component, is about half that of ($D_1 + D_2$), thus it constitutes about 7 %. As a whole, GR2 corresponds to about 25 % of the total rust.

Figure 1(d) corresponds to the spectrum of a sample close to a perforation at the height of 2 m (0.50 m above the hole). The spectrum is measured at 78 K on an enlarged scale of

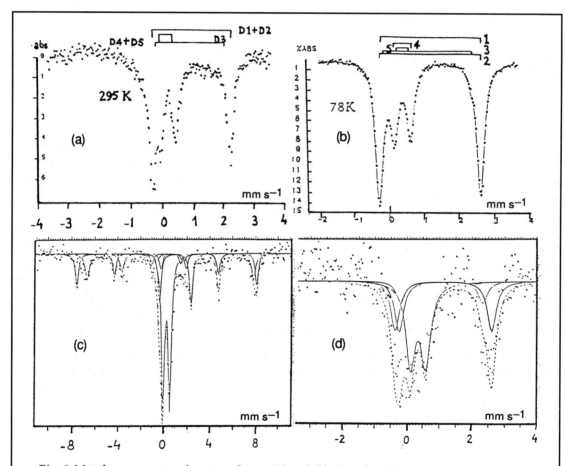

Fig. 1 Mössbauer spectra of rust products: (a) and (b). Standard Green rust 2 run at 295 K and 78 K. (c). Sample taken far from a perforation containing magnetite and green rust 2 at 295 K. (d). Sample taken close to a perforation containing essentially green rust 2 at 78 K.

± 4 mm s^{-1}. It displays the characteristics of synthetic GR2 partially oxidised in air. The product is composed of GR2 and a small amorphous paramagnetic component.

X-ray diffraction patterns of the two samples are displayed in Fig. 2. Figure 2(a) corresponds to the sample taken at 3 m height as in Fig. 1(c) and Fig. 2(b) to that taken close to the perforation at 2 m height as in Fig. 1(d). The lines of magnetite are denoted close to the perforation at 2m height as in Fig. 1(d). The lines of magnetite are denoted M whereas those of GR2 which are only visible in Fig. 2(b) are denoted RV. The amorphous nature of the paramagnetic ferric component which superimposed in the Mössbauer spectra is thus confirmed. The X-ray diffraction analysis confirms the Mössbauer spectroscopy results but only the latter detects only iron containing compounds. In Fig. 2(a), there are many lines other than those due to the rust, i.e. due to the gangue and shells. Finally, no iron sulphide component was identified within the sensitivity of the methods which have been used.

3. Conclusion

From this study of microbially induced corrosion of steel sheet piles in the harbour of Boulogne, France, we have demonstrated that there exists a comcomitance between high concentrations of sulphate reducing bacteria in the anaerobic layer of concretions on the steel surface and the presence of green rust 2 compound. Magnetite is also observed, but no sulphide is detected within the limits of the sensitivity of the method. The usual corrosion due to chloride ions must be excluded since neither green rust one nor lepidocrocite which characterise such processes are detected. A more detailed report about this study will be published later in order to try to correlate the type of rust which is obtained, the velocity of corrosion and the varieties of sulphate reducing bacteria in relation to the height from the sediment level. There is no doubt that, qualitatively, green rust 2 and microbially induced corrosion of steel by sulphate reducing bacteria are connected.

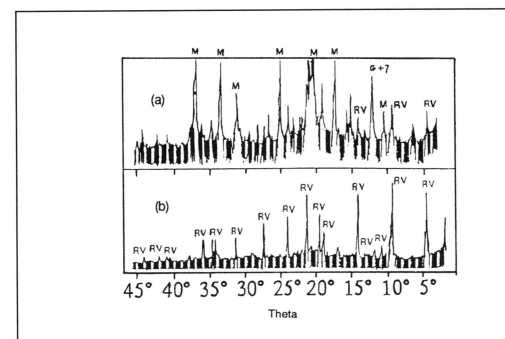

Fig. 2 X-ray diffraction patterns of the rust products: (a) Sample taken far from a perforation containing magnetite and green rust; (b) Sample taken close to a perforation containing essentially green rust 2.

Table 2 Hyperfine parameters of Mössbauer spectra, δ (mms^{-1}) = isomer shift (reference α iron), ΔE_Q (mms^{-1}) = quadrupole splitting, H (kOe) = magnetic hyperfine field, AR = relative abundance, W = full width at half maximum

	δ	ΔE_Q	AR	W	δ	ΔE_Q	H	AR	W
295K	Standard green rust 2, Fig. 1(a)				Sample at height 3 m, Fig. 1(c)				
D_{1+2}	1.13(2)	2.62(2)	4/7		1.11(4)	2.64(4)		14(2)	0.31
D_3	1.03(4)	2.26(5)	1/7		0.96(4)	2.12(4)		4(1)	0.31
D_{4+5}	0.36(2)	0.49(2)	2/7		0.38(2)	0.58(2)		40(4)	0.31
M_1					0.30(2)		486	18(2)	0.31
M_2					0.62(2)		450	19(2)	0.44
G					0.36(3)	−0.27(4)	348	5(1)	0.50
78K	Standard green rust 2, Fig. 1(b)				Sample at height 2m, Fig. 1(d)				
D_1	1.26(1)	2.96(2)	2/7		1.22(2)	2.92(10)		23(2)	0.33
D_2	1.29(1)	2.90(2)	2/7		1.28(8)	2.81(4)		23(2)	0.33
D_3	1.16(2)	2.59(2)	1/7		1.17(4)	2.51(3)		11(1)	0.33
D_4	0.46(1)	0.53(2)	1/7		0.46(3)	0.47(3)		43(3)	0.36
D_5	0.47(1)	0.37(2)	1/7						

References

1. A. A. Olowe, Ph. Bauer, J. M. Genin and J. Guezennec, Corrosion, 1989, **45**, 229-235.
2. J. A. Hardy, Journal of Applied Bacteriology, 1981, **51**, 505-516.
3. F. Bak and N. Pfennig, FEMS Microbiology Ecology, 1991, **85**, 42-52.
4. A. Girard, Thesis, Lille, 1935.
5. A. A. Olowe, Thesis, Nancy, 1988.
6. A. A. Olowe and J. M. Génin, Corrosion Science, 1991, **32**, (9), 965-984.
7. A. A. Olowe and J. M. Génin, Proc. Int. Symp. on Corrosion Science and Engineering, CEBELCOR, 1989, Report, 157, RT 297, 363-380.
8. Ph. Refait, J. M. Génin and A. A. Olowe, Role of green rust compounds in aqueous corrosion of iron in aggressive media close to marine environment, this volume, pp.167–187.
9. A. A. Olowe, J. M. Génin and Ph. Bauer, Hyperfine Interactions, 1989, **46**, 437-443.

The Role of Green Rust Compounds in Aqueous Corrosion of Iron in Aggressive Media Close to a Marine Environment

PH. REFAIT, J.-M. R. GÉNIN AND A.A. OLOWE

Laboratoire CNRS Maurice Letort (UPR 6851), Département Sciences des Matériaux, ESSTIN, Université de Nancy 1, Parc Robert Bentz, 54 500 Vandoeuvre-Nancy, France

Abstract

The oxidation of a ferrous hydroxide precipitated from a solution of ferrous salt ($FeCl_2$, $4H_2O$ or $FeSO_4$, $7H_2O$) mixed with a NaOH solution depends on the ratio R = <Fe^{2+}>/<OH^-> in the initial amounts of the reactants. For values of R > 1/2, a ferrous-ferric compound named green rust which characterises the anion, Cl^- or SO_4^{2-}, is obtained in a first stage of oxidation before the end rust product lepidocrocite. Electrochemical monitoring coupled with Mössbauer analysis leads to the chemical formula of chlorinated green rust one (GR1) as $3Fe^{(II)}(OH)_2$, $Fe^{(III)}(OH)_2Cl$, nH_2O (2 ≤ n ≤ 3) and the sulphated green rust two (GR2) as $4Fe^{(II)}(OH)_2$, $2Fe^{(III)}OOH$, $Fe^{(II)}SO_4$, mH_2O (m ≤ 4). Standard chemical potentials of green rusts and Pourbaix diagrams of iron specific to the chlorinated or sulphated aqueous medium are determined. The oxidation of $Fe^{(II)}$ to $Fe^{(III)}$ is connected with the incorporation of Cl^- or SO_4^{2-} ions in the ferrous hydroxide to produce the corresponding green rust. In particular, GR1 is shown to be an intercalation compound obtained from $Fe(OH)_2$. The oxidation of a ferrous hydroxide precipitated from a mixture of ferrous chloride and sulphate with caustic soda shows the formation of GR1 before GR2 and finally lepidocrocite. Thus GR2 can be obtained by oxidation of GR1. The role of GR1 in the pitting of steels in a marine environment and that of GR2 for microbially induced corrosion in anaerobic conditions in the presence of sulphate reducing bacteria is put forward.

1. Introduction

Chloride and sulphate anions play a predominant role in the mechanisms of corrosion of iron in neutral or basic media through intermediate compounds known as green rusts which are ferrous–ferric compounds incorporating a certain amount of these anions. 'Planar' anions, such as Cl^- or CO_3^{2-} give rise to green rust one (GR1), whereas 'tridimensional' anions, such as SO_4^{2-} give rise to green rust two (GR2). This distinction between GR1 and GR2 was established by Bernal et al. [1] on the basis of X-ray diffraction patterns. GR1 was originally discovered by Keller [2] and Yoshioka [3] more than forty years ago and GR2 by Girard [4] in 1935.

Chlorinated GR1 must obviously play an important role in the corrosion of iron and steels in a marine environment. However, even though the quantity of sulphate ions, which is present in seawater, is much smaller than that of chloride ions [5], it turns out that sulphated GR2 can play a determining role in marine corrosion. It has been shown recently [6,7] that GR2 is the product obtained by microbially induced corrosion of iron and steels and there is a concomitance between the presence of sulphate reducing bacteria and that of GR2.

In this article, we present a synthesis of the principal results obtained from the studies of the oxidation of Fe(OH)$_2$ in aqueous solution in the presence of Cl$^-$ or SO$_4^{2-}$ ions. Further details about these works can be found elsewhere [8–16].

From these results, we have tried to understand the role of the green rusts in the corrosion of iron. More specifically, we have looked for the reasons why anions which are present in the external medium are engaged in the formation of the corrosion products despite being neither oxidised nor reduced during the reaction.

2. Experimental Method

In order to study the mechanisms of oxidation of Fe$^{(II)}$ into Fe$^{(III)}$ and the formation of rusts, the following experimental method is used.

A ferrous hydroxide is precipitated from 100 ml of a solution of a ferrous salt FeX$_n$ and 100 ml of a solution of caustic soda NaOH. The oxidation of the precipitate in the solution is studied by three monitoring electrodes viz. a glass electrode for measuring the pH, a platinum electrode for measuring the electrode potential of the solution, and a saturated calomel reference electrode. These electrodes are connected to a recorder which follows the changes with time of pH and electrode potential. The temperature of the solution is kept constant by immersing the beaker in a thermostated bath. The results described here concern experiments made at 25 ± 0.5 °C. The aeration of the precipitate is assumed to be constant, since the solution was stirred with a magnetic rod rotating at a constant velocity.

If the initial NaOH concentration is fixed for a series of experiments, there will be only one unique parameter R determining the initial concentrations of the reactants. R is defined to be R = <Fe^{2+}>/<OH$^-$> where < > designates the concentration of the considered species. It is sometimes easier to use the ratio R' = <X$^{(2/n)-}$>/<OH$^-$> where X$^{(2/n)-}$ is the anion of the FeX$_n$ salt. Thus R' = nR. In the case where the hydroxide is precipitated initially by means of two ferrous salts, FeX$_n$ and FeY$_m$, one must also define a ratio which determines the proportions of the two ferrous salts, i.e. A=<X$^{(2/n)-}$>/<Y$^{(2/m)-}$>.

The results from three series of experiments will be examined. The first deals with an hydroxide precipitated from a ferrous chloride FeCl$_2$, 4H$_2$O solution with <NaOH> concentration fixed at 0.4 mol l^{-1}; the second deals with an hydroxide which is precipitated from a ferrous sulphate FeSO$_4$, 7H$_2$O solution for a concentration <NaOH> fixed at 0.2 mol l^{-1}; and the third deals with a mixture of FeCl$_2$, 4H$_2$O and FeSO$_4$, 7H$_2$O such that A = <SO$_4^{2-}$>/<Cl$^-$> = 1/8 for a concentration <NaOH> fixed at 0.2 mol l^{-1}.

It should be recalled that the value R = <Fe^{2+}>/<OH$^-$> = 1/2 corresponds, whatever the ferrous salt or the mixture of ferrous salts, to the stoichiometry of the precipitation reaction of Fe(OH)$_2$. In the case of ferrous sulphate, for R = 1/2 we have:

$$FeSO_4 + 2\,NaOH \rightarrow Fe(OH)_2 + Na_2SO_4$$

This value of R = 1/2 represents a step between the basic medium for R < 1/2 and the acidic medium for R > 1/2. In the basic medium, Fe(OH)$_2$ is obtained in the presence of an excess of caustic soda NaOH, i.e. an excess of OH$^-$ ions. The considered anion, SO$_4^{2-}$ or Cl$^-$, does not play any role in the mechanism of oxidation. In the acidic medium, Fe(OH)$_2$ is obtained in the presence of an excess of ferrous salt. Chlorinated and sulphated green rusts are thus forming.

3. Experimental Results

3.1 E_h and pH vs time curves

The E_h and pH vs time curves are presented in Fig. 1 for the chlorinated medium with ratio R = 0.80 and R = 0.54. The electrochemical potential axis is reversed to facilitate the comparison

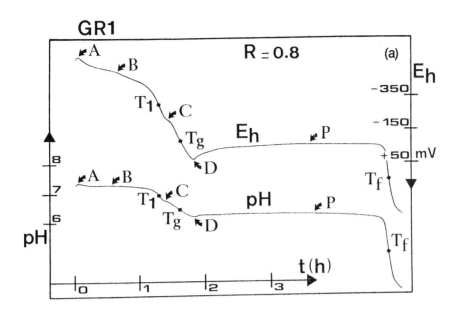

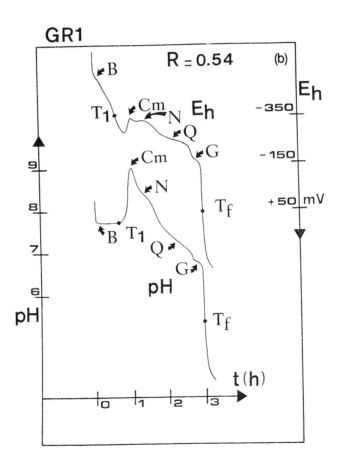

Fig. 1 Recorded E_h (1a) and pH (1b) curves with respect to time in the chlorinated medium. R = 0.8 and R = 0.54. Points A, B, C, D, P, C_m, N, Q and G are equilibrium points. T_1, T_g and T_f correspond to ends of reaction stages. Increasing value for E_h is downward.

between the E_h/time and pH/time curves. The hydrogen standard electrode is taken as the reference, i.e. $E_h = E_{calomel} + 0.245$ V

The curves obtained for R = 0.80 are typical of the reaction: Fe(OH)$_2$ → chlorinated GR1 → lepidocrocite, the γ ferric oxyhydroxide. On these curves, the inflection points T_1, T_g and T_f indicate the end of a step of the overall reaction, or, more specifically the disappearance of one reactant. T_g corresponds to the formation of GR1. The first stage of the reaction comprises two steps, one which terminates at T_1 and another one which scans from T_1 to T_g. T_1 actually corresponds to the disappearance of the initial ferrous hydroxide and the reaction between T_1 and T_g has been interpreted [10] as the transformation of nonstoichiometric GR1 into stoichiometric GR1. The second stage of reaction corresponds to the oxidation of chlorinated GR1 and T_f represents the stable end product, i.e. lepidocrocite.

When R is smaller than a value R_c, the shapes of the curves E_h and pH vs time differ. Those obtained for R = 0.54 are a good illustration. One notes that the pH/time curve displays a peak corresponding to a sharp increase of the pH, slightly after point T_1, during the second step of the formation of GR1. With the help of all the E_h and pH/time curves, one determines the step value R_c which characterises GR1, i.e. $R_c = 4/7 \approx 0.5714$ [10]. The particular points noted A, B, C, D and P on the curves obtained for R = 0.80 and B, C_m, N, Q and G on the curves obtained for R = 0.54 indicate the electrochemical equilibria between reactants and products. At these points, the electrochemical potential E_h, the pH and the activities of the ionic species match the equilibrium conditions. We have demonstrated that point B corresponds to the equilibrium between Fe(OH)$_2$ and GR1 [10], point P to that between GR1 and γFeOOH, point D to that between GR1 and the nuclei of undeveloped FeOOH and point Q to that between GR1 and hydrated magnetite of formula Fe(OH)$_2$, 2FeOOH which is suspected to precede the formation of magnetite Fe$_3$O$_4$ [9].

The E_h and pH vs time curves obtained in the sulphated medium for R = 0.625 and R = 0.5714 are presented in Fig. 2. Two main stages of oxidation are displayed, i.e. the formation of sulphated GR2 ending at T_g and the formation of the final product which proceeds from T_g to T_f. Point T_1 is absent here and the formation of GR2 only proceeds through a unique step. The pH vs time curve obtained for R = 0.5714 shows the pH peak which indicates that R is now smaller than R_c. As previously, this increase of pH takes place at the end of the formation of green rust. R_c is for GR2 equal to $7/12 \approx 0.5833$ [9]. Note that the ratios $R_c = [<Fe^{2+}>/<OH^->]_c$ are 4/7 and 7/12 for GR1 and GR2 respectively. The curves present also, as in the chlorinated medium, singular points where the electrochemical equilibrium conditions are met.

The curves of Fig. 3 (p.172) are relative to the chloro-sulphated medium such that the ratio A = $<SO_4^{2-}>/<Cl^->$ = 1/8. The cases with ratios R = $<Fe^{2+}>/<OH^->$ = 1 and 0.5714 (4/7) are presented.

The E_h and pH vs time curves obtained for R = 1 show clearly that the reaction is now made up of three major stages. Two stages are no longer solely observed as were in the chlorinated or sulphated media for the formation of the respective GR1 or GR2.

The Eh/time curve obtained for R = 0.5714 also displays three stages, although not as clearly as before. The two first stages merge in the pH/time curve where the pH stays constant. After these two stages, the pH/time curve displays the pH peak which characterises small R ratios, or more precisely ratios of R smaller than the step value R_c. Consequently, a chloro-sulphated medium can also be characterised by the value of its R_c ratio. In the case where A = 1/8, R_c is $7/12 \approx 0.5833$ [10] and has the same value as that found in the solely sulphated medium.

The meaning of the presence of this sharp increase of pH which characterises ratios R smaller than R_c has been discussed elsewhere [9, 10, 12]. It has been demonstrated [10] that the rejection of OH$^-$ ions into solution which corresponds to this increase of pH allows the totality of the initial ferrous hydroxide to be transformed into green rust, even though the quantity of ferrous salt present in solution is not sufficient.

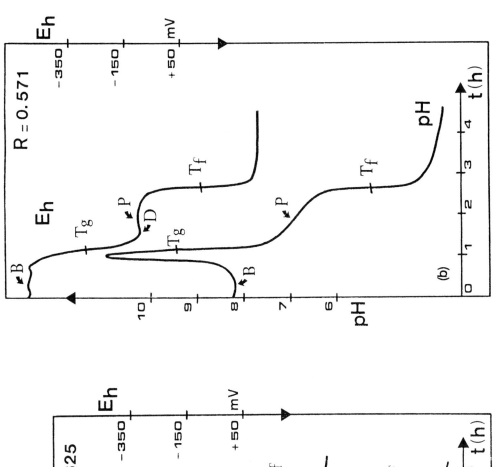
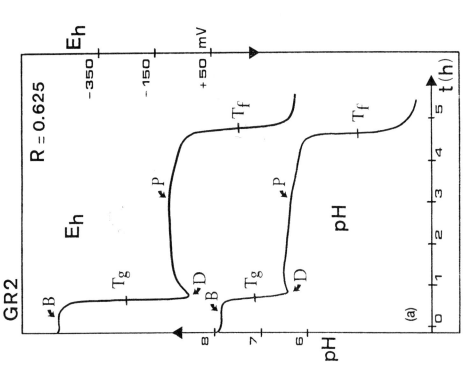

Fig. 2 Recorded E_h (2a) and pH (2b) curves with respect to time in the sulphated medium. $R = 0.625$ and $R = 0.571$. Points B, D, P are equilibrium points and T_g and T_f correspond to ends of reaction stages. Increasing value for E_h is downward.

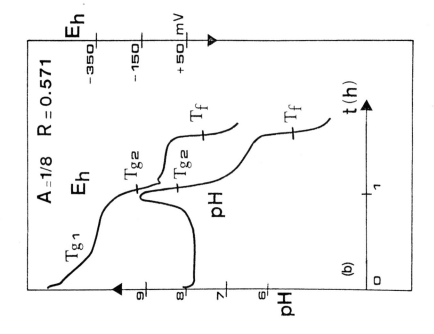
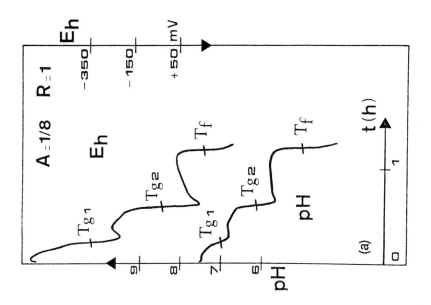

Fig. 3 Recorded E_h (3a) and pH (3b) curves with respect to time in the chloro-sulphated medium with $A = <SO_4^{2-}> / <OH^-> = 1/8$. $R = 1$ and $R = 0.571$.
Points T_{g1}, T_{g2} and T_f correspond to ends of reaction stages. Increasing value for E_h is downward.

3.2 The final products of reaction vs R curves

The final products obtained for various values of R have been analysed in the case of the chlorinated medium by X-ray diffraction [10]. These analyses only permit a semi-quantitative estimate to be made of the composition of the product. Table 1 presents the results and Fig. 4 shows the development of the composition with respect to R.

Note that lepidocrocite is obtained alone over a large domain of R ratio surrounding the characteristic ratio $R_c = 4/7$. Magnetite appears for $R = 0.55$ and becomes predominant for smaller values of R. For the highest values of R, typically for $R \geq 0.875$, the αFeOOH goethite and βFeO(OH)$_{1-x}$Cl$_x$ akaganeite appear.

In the case of the sulphated medium, we have made an analysis of the final products by X-ray diffraction and Mössbauer spectroscopy [9]. The latter method allows a precise quantitative determination of the composition of the mixture. Table 2 displays the results and Fig. 5 shows the corresponding graph.

The development of the composition of the final product with R is not very different when compared to the previous case. Around the ratio $R_c = 7/12$ a mixture of α and γFeOOH is obtained and the magnetite appears only for $R \leq 0.5625$. The only difference with the chlorinated medium is that the formation of lepidocrocite seems to be more difficult. The product obtained at R_c is the one richest in lepidocrocite (75%). When R increases, this proportion decreases, replaced proportionally by goethite. This phenomenon is also observed in chlorinated medium, but for higher R ratios. Finally, lepidocrocite is not obtained when $R < R_c$.

The end products obtained in the chloro-sulphated medium with $A = 1/8$ have been analysed by X-ray diffraction [10]. Table 3 displays the results.

It can be seen that lepidocrocite is obtained alone at the R_c ratio (0.583). γFeOOH is also predominant for the values surrounding R_c, such as 0.571, and up to high values of R such as $R = 1$. This evolution is similar to that observed in the chlorinated medium.

Table 1 Final products of the reaction in the chlorinated medium [10]

R	% Lepidocrocite	% Goethite	% Magnetite	% Akaganeite
0.51	5	2	93	0
0.515	1	13	86	0
0.525	0	15	85	0
0.54	1	14	85	0
0.5475	35	18	47	0
0.55	60	22	18	0
0.555	90	10	0	0
0.565	100	0	0	0
0.60	100	0	0	0
0.75	100	0	0	0
0.875	95	5	0	0
1.0	75	20	0	5
1.5	45	45	0	10

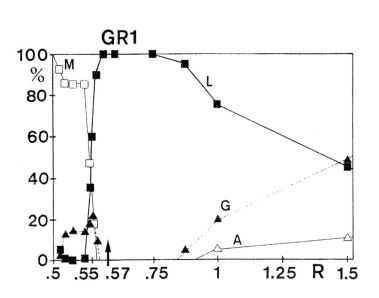

Fig. 4 Development of the proportions of the end products of oxidation vs R in the chlorinated medium.
The scales on the abscissa axis are different for R smaller or larger than 0.57.
G = goethite, L = lepidocrocite, M = magnetite and A = akaganeite.

3.3 The intermediate products of oxidation: the green rusts

Chlorinated GR1 and sulphated GR2 can be identified by their X-ray diffraction patterns [17].

In Fig. 6, the spectra of a GR1 sample obtained for R = 0.80 and of a GR2 sample obtained for R = 0.7 are displayed. The sample of GR1 had been obtained the following way: the reaction which had reached T_g is stopped and the solution is stored for four days in the absence of oxygen. After this ageing period, the precipitate is filtered under the protection of a membrane and analysed using Co K_α radiation.

The lines of GR1 are denoted by triangles with the corresponding (hkl) indices as given by Bernal *et al.* [1]. The intensities of lines (00m) are abnormally strong due to preferential orientation phenomenon of the crystallites. Besides these lines, the main lines of lepidocrocite, denoted L, are present.

The sample of GR2 was prepared differently. After an ageing period of two days, the precipitate was filtered and dried under vacuum at 30°C. Besides the lines of GR2, only those of thenardite Na_2SO_4, denoted T, are present.

The corresponding quantitative data of these two diffraction patterns and the data from ASTM cards [17] of green rusts are collected in Table 4.

The intermediate products have also been analysed in the chloro-sulphated medium when A = 1/8 and R = 1.

The spectrum of the product of the first stage of this reaction, aged one week in solution before analysis, is presented in Fig. 7(a). It includes essentially GR1, the lines of which are denoted 1 and some GR2 (denoted 2). The main lines of lepidocrocite are also present (L).

The product of the second stage has been aged and analysed in the same conditions. Its spectrum is shown in Fig. 7(b). The main lines (00m) of GR2 are distinguished. Consequently, the product of the first stage is chlorinated GR1 and the product of the second stage is sulphated

Table 2 Final products of the reaction in the sulphate medium [12]

R	% Lepidocrocite	% Goethite	% Magnetite
0.50	0	10	90
0.52	0	16	84
0.56	0	39	61
0.5625	0	66	34
0.571	0	100	0
0.583	75	25	0
0.60	55	45	0
0.70	50	50	0
1.0	32	68	0

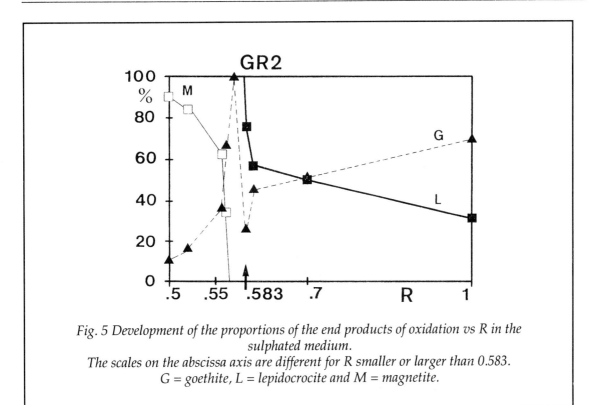

Fig. 5 Development of the proportions of the end products of oxidation vs R in the sulphated medium.
The scales on the abscissa axis are different for R smaller or larger than 0.583.
G = goethite, L = lepidocrocite and M = magnetite.

GR2. The extraneous oxidation stage which is specific of the chloro-sulphated medium is a transformation of GR1 into GR2.

Mössbauer spectroscopy allows the identification of green rusts. Spectra of GR1 obtained for R = 1 and of GR2 obtained for R = 0.7 are presented in Fig. 8. Samples are analysed at 78 K after ageing in a neutral nitrogen atmosphere to protect them against any further oxidation. Results of the computed fittings of these spectra are gathered in Table 5.

Table 3 Final products of the reaction in the chloro-sulphated medium with A = 1/8 [10]

R	% Lepidocrocite	% Goethite
0.571	100	0
0.583	100	0
0.86	80	20
1.0	60	40

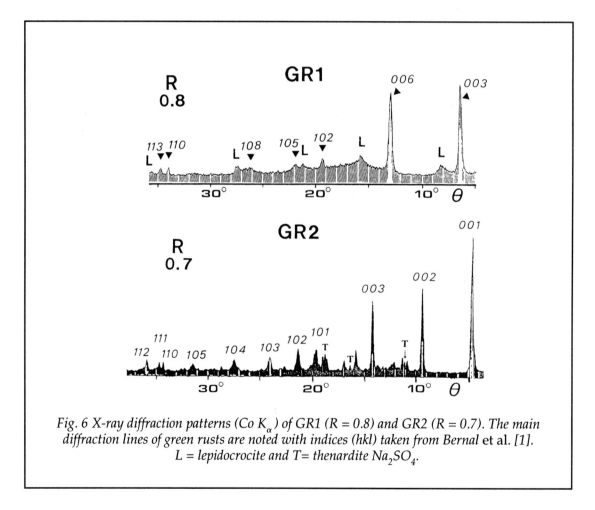

Fig. 6 X-ray diffraction patterns (Co K_α) of GR1 (R = 0.8) and GR2 (R = 0.7). The main diffraction lines of green rusts are noted with indices (hkl) taken from Bernal et al. [1]. L = lepidocrocite and T = thenardite Na_2SO_4.

Spectra at 78 K of green rusts are made of two series of paramagnetic doublets: doublets with a large quadrupole splitting QS of the order of 3 mm s^{-1} and an isomer shift, IS, of about 1.2 mm s^{-1} with respect to metallic iron and doublets with a small QS of about 0.4 mm s^{-1} and IS about 0.4 mm s^{-1}. The first series, R_1 and R_2 for GR1, D_1, D_2 and D_3 for GR2, characterise the ferrous ions whereas the second series, R_3 for GR1, D_4 and D_5 for GR2, characterise the ferric ions. The intensities of the doublets are within well defined ratios, i.e. R_1: R_2: R_3 = 2:1:1 for GR1 and D_1: D_2: D_3: D_4: D_5 = 2:2:1:1:1 for GR2 respectively.

Table 4 X-ray diffraction patterns of GR1 and GR2; data [9, 10]

GR1 R = 0.8		ASTM 13-88			GR2 R = 0.7		ASTM 13-92		
d(Å)	I/I$_1$	d(Å)	I/I$_1$	hkl(*)	d(Å)	I/I$_1$	d(Å)	I/I$_1$	hkl(*)
8.00	100	8.02	100	003	11.0	100	10.9	100	001
3.99	90	4.01	80	006	5.48	55	5.48	80	002
2.70	18	2.70	60	012	3.65	37	3.65	80	003
2.40	12	2.41	60	015	2.75	12	2.75	40	004/100
2.03	9	2.04	30	018	2.67	19	2.66	60	101
		1.81	20	10,10	2.46	23	2.46	60	102
		1.72	20	01,11	2.20	21	2.20	60	005/103
1.60	9	1.60	40	110	1.94	16	1.94	60	104
1.57	9	1.57	40	113	1.72	11	1.71	20	105
		1.54	10	10,13	1.59	9	1.59	20	110
		1.49	30	116	1.57	7	1.57	20	111
					1.52	11	1.53	20	112

* Indices given by Bernal et al. [1]

4. Interpretation of the Results

4.1 The chemical formulae of green rusts

From the results obtained by Mössbauer spectroscopy and from the value of the characteristic step ratio R_c, the chemical formulae of green rust are determined. Mössbauer spectra give the values of the Fe^{2+}/Fe^{3+} ratios of the compounds, i.e. 3:1 and 5:2 for GR1 and GR2 respectively. R_c is the ratio R which corresponds to the stoichiometry of the formation of the particularly green rust. The chemical formulae are obtained by writing the equations which describes the reaction.

Therefore, since

$$R_c = <FeCl_2>/<NaOH> = 4/7$$

for GR1:

(A) $8 FeCl_2 + 14 NaOH \rightarrow 7 Fe(OH)_2 + 14 NaCl + Fe^{2+} + 2 Cl^-$

(B) $7 Fe(OH)_2 + Fe^{2+} + 2 Cl^- + 1/2 O_2 + (2n+1) H_2O \rightarrow 2 [3Fe(OH)_2, Fe(OH)_2Cl, nH_2O]$

and the formula for GR1 is $3Fe(OH)_2, Fe(OH)_2Cl, nH_2O$.

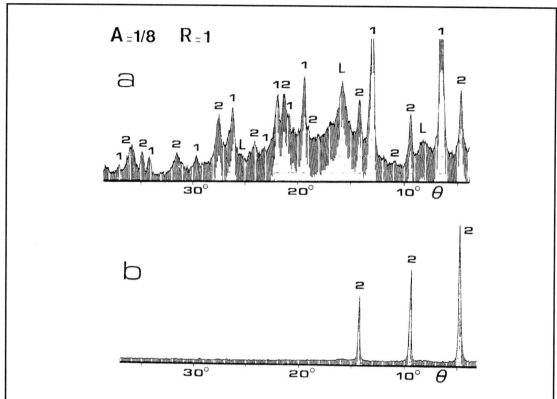

*Fig. 7 X-ray diffraction patterns (Co K_α) of the intermediate products at $A = 1/8$ and $R = 1$.
(a) Product of the first stage. (b) Product of the second stage.
1 = GR1, 2 = GR2 and L = lepidocrocite.*

Since $R_c = \langle FeSO_4\rangle / \langle NaOH\rangle = 7/12$ for GR2:

(C) $7\ FeSO_4 + 12\ NaOH \rightarrow 6\ Fe(OH)_2 + 6Na_2SO_4 + Fe^{2+} + SO_4^{2-}$

(D) $6\ Fe(OH)_2 + Fe^{2+} + SO_4^{2-} + 1/2\ O_2 + (m-1)\ H_2O \rightarrow 4Fe(OH)_2, 2FeOOH, FeSO_4, mH_2O$

and the formula for GR2 is $4Fe(OH)_2, 2FeOOH, FeSO_4, mH_2O$.

These two chemical formulae have a most interesting common property: the number of negative charges carried by the foreign anion is equal to the number of Fe^{3+} ions, i.e. to the excess of positive charges carried by Fe^{3+} ions with respect to Fe^{2+} ions. In other words, $Fe^{3+}/Cl^- = 1$ for GR1 and $Fe^{3+}/SO_4^{2-} = 2$ for GR2.

Note also that the oxidation degree of GR2 is 2.286, greater than that of GR1, which is 2.250. Therefore, GR2 can be obtained by oxidation of GR1, as observed in a chloro-sulphated medium for $A = 1/8$, whereas the reverse is impossible.

4.2 Pourbaix diagrams of green rusts

Knowing the chemical formula of a green rust, one can determine its standard chemical potential $\mu°$ from the values of E_h and pH which are measured at the equilibrium points where the corresponding green rust is observed.

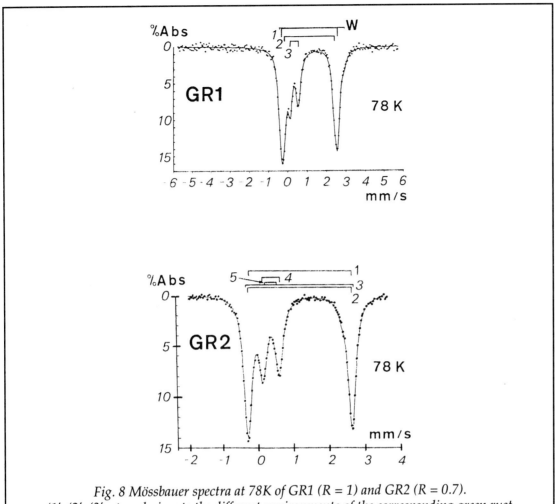

*Fig. 8 Mössbauer spectra at 78K of GR1 (R = 1) and GR2 (R = 0.7).
'1', '2', '3', etc... designate the different environments of the corresponding green rust
and W designates Fe^{2+} ions trapped in ice.*

Table 5 Mössbauer spectra of GR1 (R = 1) and GR2 (R = 0.7) [11, 14]

GR1 Site	%	IS (mm/s)	QS (mm/s)	GR2 Site	%	IS (mm/s)	QS (mm/s)
R1	1/2	1.26	2.88	D1	2/7	1.256	2.88
R2	1/4	1.25	2.60	D2	2/7	1.237	2.915
R3	1/4	0.47	0.41	D3	1/7	1.224	3.01
				D4	1/7	0.434	0.515
				D5	1/7	0.412	0.35

The reference for all isomer shifts is metallic α-iron.

For GR1, the chemical potential is computed to be [10]:

$$\mu°[3Fe(OH)_2, Fe(OH)_2Cl] = -509\,500 \pm 500 \text{ cal mol}^{-1},$$

from the equilibria Fe(OH)$_2$ <—> GR1 (point B), GR1 <—> γFeOOH (point P) and finally GR1 <—> hydrated magnetite (point Q).

For GR2, the chemical potential is computed to be from the Fe(OH)$_2$ <—> GR2 equilibrium [9, 13]:

$$\mu°[4Fe(OH)_2, 2FeOOH, FeSO_4] = -902\,890 \text{ cal mol}^{-1}$$

but more precisely, if using the activity of SO_4^{2-} and not its concentration, this value is found to be [10]:

$$\mu°[4Fe(OH)_2, 2FeOOH, FeSO_4] = -904\,100 \pm 350 \text{ cal mol}^{-1}.$$

The development of the final products of oxidation of a ferrous hydroxide with R in the presence of Cl^- or SO_4^{2-} ions (Figs. 4 and 5) shows that, whatever the considered green rust, lepidocrocite γFeOOH is in principle the product of the reaction. When R is equal to R_c, the reactants Fe(OH)$_2$ and FeCl$_2$ or FeSO$_4$ are entirely consumed during the formation of the green rust which is then oxidised to lepidocrocite. When R is different from R_c, the dissolved species can favour the production of another compound. The Fe^{2+} and OH^- ions are known to favour the formation of αFeOOH, goethite [18–22]. On the other hand, when R is much lower than R_c, the quantity of ferrous salt becomes insufficient to transform all the initial ferrous hydroxide. New mechanisms of oxidation appear and give rise to magnetite.

However, it is clear that the transformation Fe(OH)$_2$ → GR1 → γFeOOH characterises the oxidation of iron in a medium which contains foreign anions such as Cl^- or SO_4^{2-}. Therefore, the corrosion of iron in such a medium must be illustrated by a specific Pourbaix diagram which includes the corresponding green rust and lepidocrocite, γFeOOH. Since we have determined the chemical formulae and potentials of GR1 and GR2 such diagrams can now be presented.

The diagram concerning a chlorinated medium is drawn in Fig. 9 for a temperature of 25°C and an activity of Cl^- ions of 0.35 mol l^{-1} corresponding to that of seawater [5]. On this diagram GR1 is designated by its condensed formula Fe$_4$(OH)$_8$Cl. We have not considered here the existence of chlorinated ferrous hydroxides, such as the hydroxide 2 proposed by Rezel [8, 11] with a formula 2Fe(OH)$_2$, FeOHCl. The study of these hydroxides and of the mechanisms of their oxidation is in progress but is far from finished.

The diagram concerning a sulphated medium is drawn in Fig. 10 for a temperature of 25°C and an activity of SO_4^{2-} or HSO_4^- ions of 0.1 mol l^{-1}. The notation 'BS' designates the basic sulphate, with the variable formula (1-β)Fe(OH)$_2$, βFeSO$_4$ ($0 \leq \beta \leq 1$) which is obtained in place of Fe(OH)$_2$ when the ratio R is high, i.e. greater than 0.625 [9, 13]. The construction of these diagrams is detailed elsewhere [9, 10, 13].

However, let us go back to some specific points, to make more precise the connections which exist between these diagrams and the measurements of electrochemical potential and pH. Considering first the Pourbaix diagram of chlorinated GR1, the equilibrium straight line (7) describes the equilibrium between Fe(OH)$_2$ and GR1 which corresponds to points B on the E_h and pH/time curves. This line is horizontal, and so E_h does not depend on pH. This follows from the chemical formula of GR1 which contains one Fe^{3+} ion for one Cl^- ion.

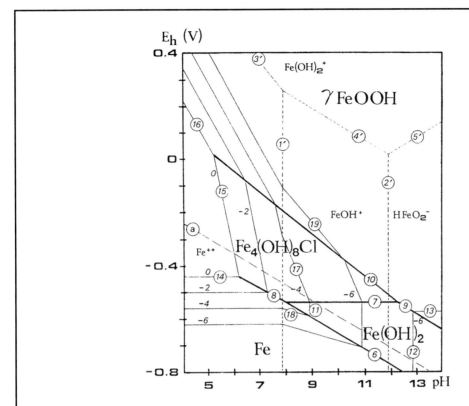

Fig. 9 E_h-pH equilibrium diagram at 25°C and activity $[Cl^-] = 0.35$ mol l^{-1} for the system Fe-GR1-γFeOOH-H$_2$O-Cl$^-$.
GR1 = Fe$_4$(OH)$_8$Cl;
$\mu°$(GR1) = –509 500 cal mol^{-1}.

Therefore one can write:

(7a) $4 Fe(OH)_2 + Cl^- \rightarrow Fe_4(OH)_8Cl + e^-$,

giving the line:

(7b) $E_h = E° - 0.0591 \log [Cl^-]$

The E_h potentials measured at points B meet this equation (7b).
Between R = 0.513 where E_h(B) = –0.527 V and R = 0.875 where E_h(B) = –0.526 V, i.e. where E_h is constant, a regular decrease of pH from 9.15 to 7.31 is observed [10]. It matches the fact that E_h at the Fe(OH)$_2$ <—> GR1 equilibrium is independent of the pH and consequently that equation (7a) is correct, i.e. that GR1 contains one Fe^{3+} ion for one Cl$^-$ ion. At R = 1.0, one observes that E_h(B) varies with the pH. It is –0.502 V at R = 1.25 for a pH of 7.07 and –0.454 V at R = 1.5 for a pH of 6.86 [10]. GR1 is still involved in this equilibrium, since points D and P are still relative to the equilibrium GR1 <—> FeOOH for these values of R. Therefore, the initial hydroxide is no longer Fe(OH)$_2$ but a chlorinated hydroxide Fe(OH)$_{2-x}$Cl$_x$. Since some additional information about these hydroxides is still needed they are not reported in Fig. 9.

Let us consider now the diagram concerning GR2 in Fig. 10. The line (7) is also the equilibrium between Fe(OH)$_2$ and GR2. Again the line is horizontal since GR2 contains the same number of Fe^{3+} ions and of negative charges carried by the foreign anions SO$_4^{2-}$.

(7c) $7\,Fe(OH)_2 + SO_4^{2-} \rightarrow 4Fe(OH)_2, 2FeOOH, FeSO_4 + 2\,H_2O + 2\,e^-$

(7d) $E_h = E° - 0.2955 \log[SO_4^{2-}]$

Therefore, between R = 0.52 when E_h is measured to be –0.497 V and R = 0.625 when E_h is –0.495 V, E_h is indeed constant and the pH decreases regularly from 9.93 to 7.82 [9,13].

As in the case of the chlorinated medium, the potential E_h of the equilibrium hydroxide <–> green rust varies with the pH above a given value of R (R > 0.625). It is then demonstrated [9,13] that the equilibrium takes place between GR2 and a basic sulphate (BS) with formula $(1-\beta)Fe(OH)_2, \beta FeSO_4$. The variation of β is continuous between 0 and 1 and its value increases with R. The equilibrium BS <—> GR2 is described by a portion of a parabola (8), the slope of which goes from that of the equilibrium Fe(OH)$_2$ <—> GR2, i.e. zero, to that of the equilibrium FeSO$_4$ <—> GR2.

Let us consider now the equilibria GR1 <—> γFeOOH and GR2 <—> γFeOOH which correspond to line (10) on both diagrams. Point P is the corresponding experimental point observed on the E_h and pH vs time curves (see Figs. 1 and 2). But during the transformation of a green rust into ferric oxyhydroxide there exists a second equilibrium point at the beginning of the transformation, denoted as D in both chlorinated and sulphated media. Since this point is always present when lepidocrocite is formed, whether it is obtained alone or in the presence

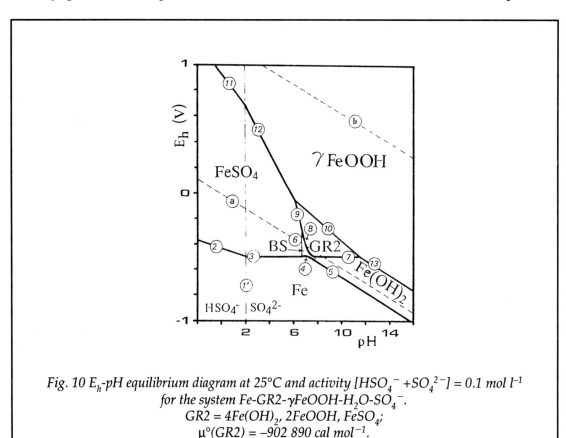

Fig. 10 E_h-pH equilibrium diagram at 25°C and activity $[HSO_4^- + SO_4^{2-}] = 0.1$ mol l^{-1} for the system Fe-GR2-γFeOOH-H$_2$O-SO$_4^-$.
GR2 = 4Fe(OH)$_2$, 2FeOOH, FeSO$_4$;
μ°(GR2) = –902 890 cal mol^{-1}.

of some goethite, it is clear that the corresponding equilibrium is connected to the formation of lepidocrocite, or, more specifically, constitutes a step in its formation with no connection with the formation of any other FeOOH phase. We have assumed [10] that it is an equilibrium between green rust and the undeveloped nuclei of γFeOOH. The standard chemical potential $\mu°$ of these nuclei is obtained from the E_h and pH values measured at points D. We have obtained [9, 10, 13]:

$$\mu°(FeOOH) \approx -111\,000 \text{ cal mol}^{-1},$$

i.e. a value which is higher than that of the chemical potential of γFeOOH, which is close to $-112\,100$ cal mol^{-1} [23].

5. Discussion

Most of the authors [24-26] who studied green rust 1 formed in the presence of carbonate ions proposed that this compound would be isomorphous with pyroaurite, a mineral with the formula $Mg^{(II)}_6Fe^{(III)}_2(OH)_{16}CO_3,4H_2O$. We have recently shown [10,16] that the crystallographic structure of chlorinated GR1 also has a strong relation with that proposed for pyroaurite even though there is a slight departure. Moreover, it will be consistent with the experimental information we discussed previously and a structural model for the oxidation of $Fe(OH)_2$ will be put forward.

The pyroaurite structure as determined from the works of Ingram and Taylor [27] and Allmann [28, 29], can be described briefly. It is rhombohedral with space group R3m. Described in the conventional hexagonal cell the parameters are a = 3.109Å and c = 23.412 Å. It is built up by a stacking of hydroxide layers with formula $[Mg^{(II)}_6Fe^{(III)}_2(OH)_{16}]^{2+}$, which have a positive charge due to the presence of trivalent cations. Between two layers of this type lies an interlayer negatively charged, composed of CO_3^{2-} ions and water molecules.

The sequence of stacking of layers is AcB i BaC j CbA k where A, B, C designate the planes of OH$^-$ ions, a, b, c the planes of metallic cations and i, j, k the interlayers. One unit cell is composed of three elementary strata of the type (AcB i). The Mg^{2+} and Fe^{3+} cations occupy all the possible octahedral sites of the hydroxide layer where they are assumed to be randomly distributed. The oxygen atoms of the CO_3^{2-} ions and H_2O molecules are also randomly distributed as long as three atoms of the same CO_3^{2-} ion are consistent with the ion geometry.

It is possible to adapt such a model to the case of ferrous–ferric hydroxides containing various anions such as green rusts. One requisite must however be fulfilled: the negative charges of the interlayers must match the positive charges of the hydroxide layers, i.e. the excess positive charges carried by the Fe^{3+} ions are counterbalanced by the negative charges carried by the foreign anions. This is the case for GR1 when one Fe^{3+} ion is counterbalanced by one Cl$^-$ ion and for GR2 when two Fe^{3+} ions are counterbalanced by one SO_4^{2-} ion.

It is of interest to compare the structures of $Fe(OH)_2$ and green rust. The sequence of stacking for the two structures is represented schematically in Fig. 11, taking as an example the chlorinated GR1 with the structure we proposed [10,16].

The sequence of stacking for the OH$^-$ ions in the ferrous hydroxide is ABA, i.e. that of a hcp sublattice. The Fe^{2+} ions occupy all the octahedral sites in between the layers A and B of OH$^-$ ions leaving empty the sites between B and A. One can write the sequence as AcBA where c represent the octahedral sites occupied by Fe^{2+} ions.

During the oxychlorination which leads to GR1, Cl$^-$ ions are inserted in the empty octahedra and the upper plane A goes into position B by slip, giving a new stacking, i.e. the stacking AcBAcB becomes AcB i BaC where i represents the inserted Cl$^-$ ions and water molecules. This process can be progressive up to the moment when all possible insertions are

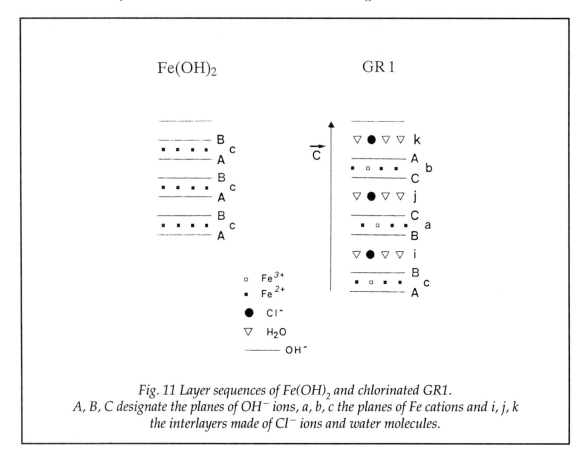

*Fig. 11 Layer sequences of $Fe(OH)_2$ and chlorinated GR1.
A, B, C designate the planes of OH^- ions, a, b, c the planes of Fe cations and i, j, k
the interlayers made of Cl^- ions and water molecules.*

accomplished. Therefore the maximum occurs when AcBAcBAcBA... gives the sequence AcB i BaC j CbA k A.... This is the corresponding stoichiometric GR1 of the type proposed for pyroaurite. Lattice parameters of chlorinated GR1 in the conventional hexagonal cell are determined to be a = 3.20Å and c = 24.00Å from the sample obtained at R = 0.8 (Table 5).

One must recall that the interlayers i, j, k are of the form $[Cl^-, nH_2O]$ with $2 < n \leq 3$, i.e. negatively charged. The Fe^{2+} ions which are originally dissolved from the ferrous chloride during the formation of GR1 (equation (B)) stay in the solution allowing only the Cl^- ions to get into the solid compound. Therefore, during the insertion of Cl^- ions, there will be a rejection of a negative charge from the $Fe(OH)_2$ precipitate. This charge is carried by an electron which comes from the oxidation of one Fe^{2+} ion of hydroxide into one Fe^{3+} ion. This electron is accepted by the dissolved oxygen in the solution. The reduction of oxygen molecules O_2 yields the OH^- ions which will probably contribute with the Fe^{2+} ions in solution to the growth of $Fe(OH)_2$ precipitates. This mechanism is illustrated in Fig. 12.

This model allows one to understand the role played by the green rusts in the oxidation of iron. The Cl^- ions, and any other foreign anion in general, induce the transformation of Fe^{2+} ions into Fe^{3+} ions by their insertion within the structure of ferrous hydroxide. They play a key role in the oxidation process.

A similar conclusion can be drawn for green rust 2 where SO_4^{2-} ions are involved. The difference in the X-ray diffraction pattern between GR1 and GR2 should be recalled, this difference being at the origin of their terminology. The structure of GR2 is not as closely related to pyroaurite as GR1 even though one deals with the insertion of SO_4^{2-} ions. But since SO_4^{2-} is a tridimensional ion, its incorporation within $Fe(OH)_2$ is not as simple as described for planar ions. The proposed structure would be AcBaC i A or AcBcA i A [9], but this should be

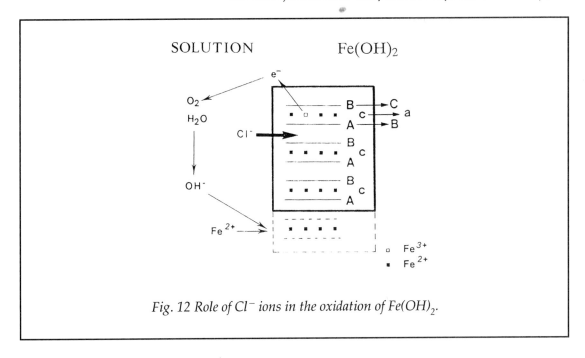

Fig. 12 Role of Cl^- ions in the oxidation of $Fe(OH)_2$.

confirmed. Anyhow, the existence of a transformation $Fe(OH)_2 \to GR1 \to GR2$, as observed in the chloro-sulphated medium, infers that Cl^- planar ions are incorporated before SO_4^{2-} tridimensional ions.

A last remark must be made with reference to the mechanism of corrosion through the formation of green rusts. The ions Cl^- or SO_4^{2-} which are incorporated in the green rusts escape later on to the external medium. For instance the reaction of oxidation of GR1 is written in the following way:

(B) $7\ Fe(OH)_2 + Fe^{2+} + 2\ Cl^- + 1/2\ O_2 + (2n+1)\ H_2O \to 2\ [3Fe(OH)_2, Fe(OH)_2Cl, nH_2O]$

(E) $3Fe(OH)_2, Fe(OH)_2Cl, nH_2O + 3/4 O_2 \to 4\gamma FeOOH + H^+ + Cl^- + (2n+3)/2\ H_2O,$

where $2 \le n \le 3$.

Equation (E) indicates that the oxidation of GR1 releases some hydrochloric acid and that this acid is formed at the GR1/γFeOOH interface, i.e. within the rust layer. It is consequently very likely that this corrosion of iron through the intermediate of chlorinated GR1 is associated with the pitting which characterises the corrosion of steels in chlorinated aqueous media. At the place where the corrosion of iron starts, the formation of GR1 followed by its oxidation to γFeOOH is accompanied by a localised acidification which provides a marked aggressivity in the presence of the released hydrochloric acid.

6. Conclusion

The chemical formula of a green rust has been determined in two ways. The Mössbauer analysis of the compound allows the determination of the Fe^{3+}/Fe^{2+} ratio. The study of the oxidation of the precipitated ferrous hydroxide from a FeX_n ferrous salt mixed with caustic soda NaOH, with respect to the ratio R of the concentrations $<FeX_n>/<NaOH>$ allows—through the determination of the critical ratio R_c, which corresponds to the stoichiometry of the formation of the green rust—the X/Fe ratio which characterises the compound to be obtained.

The formula of chlorinated GR1 is thus $3Fe(OH)_2, Fe(OH)_2Cl, nH_2O$ with $2 \leq n \leq 3$ and that of sulphated GR2 is $4Fe(OH)_2, 2FeOOH, FeSO_4, nH_2O$ with $n \leq 4$.

The Fe^{3+}/Fe^{2+} ratios of GR1 and GR2 are 1/3 and 2/5 respectively. Since the degree of oxidation of GR2 is larger than that of GR1, it is possible to transform GR1 into GR2 by oxidation. This is observed during the oxidation of ferrous hydroxide in the simultaneous presence of Cl^- and SO_4^{2-} when the ratio $A = <SO_4^{2-}>/<Cl^->$ is small, e.g. $A = 1/8$. Green rusts have some common characteristics: the negative charges carried by the anion being considered are balanced by the positive charges carried by oxidation of Fe^{2+} into Fe^{3+}. Thus GR1 contains one Cl^- ion for one Fe^{3+} ion and GR2 one SO_4^{2-} ion for two Fe^{3+} ions. The integration into the hydroxide of the Cl^- or SO_4^{2-} anions is connected with the oxidation, i.e. with the escape of the electrons to the dissolved oxygen, and we propose that this is the driving process stressing the role of green rust in the oxidation mechanism. Moreover, GR1 formed with planar anions such as Cl^- or CO_3^{2-} has a structure close to that of pyroaurite and can be considered as intercalation compounds of $Fe(OH)_2$. The stacking AcBA... in $Fe(OH)_2$ where A and B designate the OH^- planes and c the Fe^{2+} planes becomes AcBiB... for the GR1, where i is the interlayer which contains the considered anions and the water molecules, assuming that the intercalation is responsible for the slip of the layer from position A to position B.

To illustrate the role of green rusts in the corrosion of iron, the Pourbaix diagrams of iron in chlorinated and sulphated aqueous media, especially for chloride concentrations close to those found in seawater, have been drawn in the presence of the rust product which is formed by the further oxidation of green rusts, i.e. lepidocrocite. The pitting of steel is attributed to the process described here for chlorinated media and the formation of sulphated GR2 seems to be the process observed when the presence of sulphate reducing bacteria in microbially induced corrosion is active in anaerobic conditions, as is the case in marine sediments or close to the low tide level when protective layers are formed by fouling.

References

1. J. D. Bernal, D. R. Dasgupta and A. L. Mackay, Clay Min. Bull., 1959, **4**, 15-30.
2. G. Keller, Thesis, Bern, 1948.
3. H. Yoshioka, Kagaku, 1948, **18**, (9), 413.
4. A. Girard, Thesis, Lille, 1935.
5. R. M. Garrels and M. E. Thompson, Amer. J. Sci., 1962, **260**, 57-66.
6. A. A. Olowe, Ph. Bauer, J. M. R. Génin and J. Guézennec, NACE Corrosion, 1989, 45, (3), 229-235.
7. J.-M. R. Génin, A. A. Olowe, B. Resiak, N. D. Benbouzid-Rollet, M. Confente and D. Prieur, Identification of sulfated green rust two compound as a result of microbially induced corrosion of steel sheet piles in a harbour, this volume, pp.162–166.
8. D. Rezel, Thesis, Nancy, 1988.
9. A. A. Olowe, Thesis, Nancy, 1988.
10. Ph. Refait, Thesis, Nancy, 1991.
11. J. M. R. Génin, D. Rezel, Ph. Bauer, A. A. Olowe and A. Béral, Electrochem. Meth. in Corr. Res., Mat. Sci. For., 1986, **8**, 477-490.
12. A. A. Olowe and J. M. R. Génin, Corr. Sci., 1991, **32**, 965-984.
13. A. A. Olowe and J. M. R. Génin, Proc. Int. Symp. Corros. Sci. Engng., CEBELCOR RT297, 1989, 363-380.
14. A. A. Olowe, J. M. R. Génin and Ph. Bauer, Hyp. Int., 1988, **41**, 501-504.
15. A. A. Olowe and J. M. R. Génin, Hyp. Int., 1990, **57**, 2029-2036.

16. Ph. Refait, D. Rezel, A. A. Olowe and J. M. R. Génin, 'Mössbauer effect study and crystallographic structure of chlorinated green rust one compounds', ICAME '91, Nanjing, Hyp. Int., 1991, **69**, 839-842.
17. ASTM cards nos. 13-88 (GR1) and 13-92 (GR2).
18. R. Derie and M. Ghodsi, Ind. Chim. Belg., 1972, **37**, 731-740.
19. J. Detournay, M. Ghodsi and R. Derie, Ind. Chem. Belg., 1974, **39**, 695-701.
20. P. Schindler, W. Michaelis and W. Feitknecht, Helv. Chim. Acta., 1963, **46**, 444-449.
21. G. W. Van Oosterhout, J. Inorg. Nucl. Chem., 1967, **29**, 1235-1238.
22. M. François and J. P. Martiny, Institut de Chim. Ind., U.L.B., Rapport interne, 1972.
23. H. Schwartz, Werkst. und Korros., 1972, 648.
24. P. P. Stampfl, Corros. Sci., 1969, **9**, 185-187.
25. G. W. Brindley and D. L. Bish, Nature, 1976, **263**, 353.
26. R. M. Taylor, Clay Miner., 1980, **15**, 369-382.
27. L. Ingram and H. F. W. Taylor, Min. Mag., 1967, **36**, 465.
28. R. Allmann, Acta Cryst., 1968, **B24**, 927-977.
29. R. Allmann, Neues Jahrb. Min., Monatsh, 1969, 552.

PROTECTION

16

The Cathodic Protection of Steel for Offshore Platforms in Polluted Seawater

D. Yuan-long, F. Chao, L. Zheng and W. Wei

Institute of Corrosion and Protection of Metals, Chinese Academy of Sciences, Shenyang 110015, China

Abstract

The presence of H_2S or other pollution of seawater next to the surface of offshore platform joints, has led to a study of the optimum protection potential ranges for the cathodic protection of the steel for offshore platform in clean, plain and H_2S-containing polluted seawater. The results are discussed from the point of view of protection efficiency or the low stress brittle fracture of the steel resulting from hydrogen induced cracking.

1. Introduction

Cathodic protection is a necessary and most important measure in controlling the corrosion of steel for offshore platforms in seawater. There are some conventional criteria relating to cathodic protection for offshore platforms [1, 2], but all of them are based on clean seawater and protection efficiency is the main factor considered. At an estuary or harbour, seawater will be polluted to various extents. For example, in the case of the estuary of the Mississippi, the hydrogen sulphide content in the seawater fluctuates between 50 and 100 ppm or higher [3]. Furthermore, for geometrical reasons, fouling organisms may preferentially adhere to sheltered regions of the platform joints. As a result, these regions will experience localised gathering of drainage and rubbish from the platform together with H_2S pollution. Practical measurement shows that, even for new platforms and those up to four years old, the H_2S content may reach 20 to 40 ppm in the local environment next to the surface of the platform joints [4]. In addition, in line with the conventional point of view held by many designers of offshore platforms, the joints may be protected cathodically at a very negative potential, so long as hydrogen evolution can be avoided [5]. The steel for offshore platform joints is sensitive to hydrogen-induced cracking (HIC) [6]. Loading stress, residual stress and stochastic stress are always concentrated on the joint. Therefore, HIC susceptibility of the joint steel will be greatly increased under the synergistic effect of cathodic protection and H_2S-containing media. In this paper, the optimum protection potential ranges of cathodic protection of the steel for offshore platform joints in clean and H_2S-containing seawater are studied and discussed from the point of view either of the protection efficiency or the low stress brittle fracture of the joint steel caused by HIC.

2. Experimental

2.1 Test material and specimen

The steel for an offshore platform joint (JIS SM50B-ZC120, Table 1), including the matrix and weldment, was used. For measuring and evaluating the HIC susceptibility, dumbell-shaped specimens (5.00 ± 0.01 mm in diameter and 25.00 ± 0.05 mm in length for test section) are used.

Table 1 Chemical composition of JIS SM50B-ZC120 steel

C	Mn	P	S	Si	Ni	Cr	Mo	Cu	V
0.18	1.42	0.017	0.004	0.25	0.05	0.05	0.05	0.15	0.05

2.2 Experimental

(1) Corrosion testing in clean seawater. Based on these tests, cathodic protection efficiencies $\eta\%$ are calculated.

(2) Slow strain rate tensile test (SSRT). Based on the SSRT, the Index of Hydrogen Embrittlement F% is calculated as follows:

$$F\% = (\psi_0 - \psi)/\psi_0 \times 100\%$$

Where ψ_0 and ψ are the Reduction of Area (fracture shrinkage) respectively in the air and in the clean or H_2S polluted seawater.

It is generally acknowledged [7, 8] that, when steel yields, if F% is larger than 35%, the system is sensitive to HIC; if it is lower than 25%, the system is free from any hazard of HIC; and if it is between 25% and 35%, the system will be in danger of latent HIC.

(3) Fractography. The morphology of the fracture of the tensile specimen is observed using SEM.

Test machine details: Model HYS-5 SSRT machine with potential controlled by potentiostat (Hokuto Denko Model HA-501).

Test solutions:

- Artificial/plain seawater, ASTM D1141-52, pH 8.2;
- Artificial seawater with H_2S of different concentrations, de-aerated by pure nitrogen (N_2+Ar ≥ 99.999%) previously deoxygenated by reaction with an active copper catalyst.

Test temperature: 20°C.
Reference electrode: Saturated calomel electrode (SCE).

3. Results and Discussion

3.1 The effect of the cathodic protection potential (E) on the Index of the Hydrogen Embrittlement (F%) and on the Cathodic Protection Efficiency ($\eta\%$)

The effect of E on the F% and on the $\eta\%$ of the steel for offshore platform joints (weldment specimen) in clean seawater is shown in Fig.1.

It is obvious that, when E is shifted negatively, the $\eta\%$ will increase greatly; but at the same time, the steel will be susceptible to hydrogen embrittlement when E becomes more negative than –950 mV (F% larger than 25%); and when it is more negative than –1000 mV, a hazard failure of HIC will be present.

Observation of the fractographs by SEM also proves the above results. Even in clean seawater, it is advisable to take the upper limit of the protection potential at the joints, i.e. as more negative than –950 mV, and particularly not more than –1000 mV. Otherwise, when the stress becomes higher than the yield point loaded on the joint, it will be dangerous or sensitive to HIC.

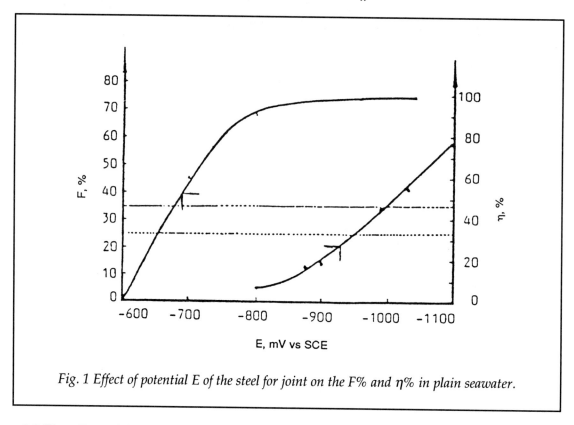

Fig. 1 *Effect of potential E of the steel for joint on the F% and η% in plain seawater.*

3.2 The effect of the H_2S concentration in artificial seawater on E and F%

The effect of the H_2S content in seawater on the E and F% of the steel for offshore platform joint is shown on Fig. 2.

It can be seen that the higher the H_2S content in seawater, the more positive the protection potential above which the steel will be free from any hazard of HIC and below which the steel will be sensitive to HIC when yielding.

Apparently, these results only reveal the hazard of HIC of the joint steel in H_2S-polluted seawater. In practice, the fracture of the joint would take place only when the joint steel is yielded or locally yielded by stress concentration.

It is clear that local H_2S pollution of the seawater next to the surface of the offshore platform joints seems to be almost inevitable, if the fouling organisms and/or sulphate-reducing bacteria are present on these sheltered and de-aerated zones. Accordingly, the periodic removal of the fouling organisms from these sites is likely to be beneficial for the safety control of the platform. In addition, the point of view, held by many designers of offshore platforms, that the more negative the cathodic protection potential of the offshore platform is, the better—so long as hydrogen evolution is avoided—needs to be reconsidered. In fact, the environmentally assisted fracture HIC of the joint steel may occur at a protection potential more positive than the above negative boundary of the potential, not only in clean seawater, but especially in H_2S polluted, or locally polluted, seawater next to the surface of the joints.

4. Suggestions

For both safety control, and for efficiency of protection, the HIC susceptibility of the steel for offshore platform joints in clean and in H_2S-polluted seawater should be reduced as low as possible. Therefore, the following suggestions appear to be of value in improving the design of cathodic protection systems for offshore platforms:

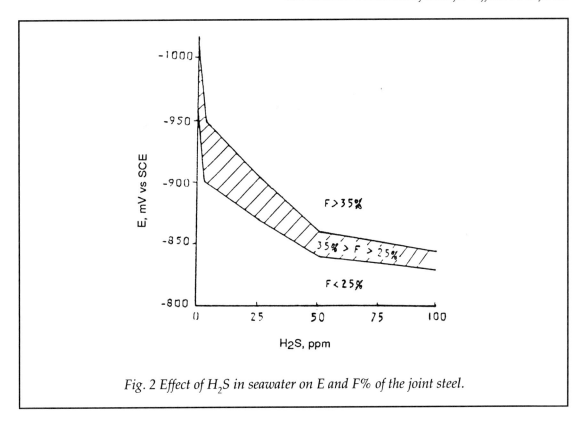

Fig. 2 Effect of H_2S in seawater on E and F% of the joint steel.

(1) In clean seawater, the cathodic protection potential needs to be maintained cathodically between –780 and –950 mV vs SCE or Ag/AgCl/seawater reference electrode. Under these conditions, the cathodic protection efficiency will still be higher than 90% and free from any hazard caused by HIC.

(2) In H_2S-polluted sea water, the optimum protection potential will change with the H_2S content of the seawater. The potential range between –780 to –830 mV is likely to be safe for many H_2S-polluted situations, provided the H_2S content is higher than 50 ppm in seawater. When the H_2S content in the seawater is decreased, the potential range might be extended to rather negative values.

(3) The periodic removal of the fouling organisms from the joints will be beneficial in avoiding the local pollution on the joint zone and therefore beneficial to the safety control of the offshore platform.

References

1. Control of corrosion on steel, fixed offshore platforms associated with petroleum production. Recommended practice, NACE Standard RP-01-76, 1976.
2. The continental shelf — Main acts, regulations and guidelines, issued by Norwegian Authorities, Norwegian Petroleum Directorate, 1983.
3. 'Chemical Engineering Progress Symposium Series', 1969, No. 97; 'Water', 1968, Vol. 65; or cf. 'Translating Collection on Geological Environmental Pollution' — Water Pollution, 1973, 9, Scientific & Technical Reference Publishing House.
4. Du Yuan-long et al., Underwater measurements by the probe of HIC susceptibility of the U-A/B offshore platform, Zengbei Oil Field, Bohai, China, Unpublished, 1988.

5. Du Yuan-long *et al.*, On the criteria of cathodic protection of steel for offshore platform. Asia Corrosion '88 International Conference, Singapore, 1988.

6. Wang Zhounguang *et al.*, Acoustic emission monitoring the fatigue crack growth in SM50B-ZC steel for the structure of offshore platform, Proc. Int. Conf. on Evaluation of Materials Performance in Severe Environments, Kobe, Japan, 1989.

7. Bastien *et al.*, Revue de Metallurgie, Memories, 1958, **55**, 310-312.

8. R. N. Tuttle and R. D. Kane (Co-edited), H_2S corrosion in oil & gas production — A compilation of classic papers, published by NACE, 1981.

Thermal Sprayed Aluminium Coatings in Seawater with and without Cathodic Protection

P. O. GARTLAND AND T. G. EGGEN

SINTEF Corrosion Center, N-7034 Trondheim, Norway

Abstract

In the last few decades the use of thermal sprayed aluminium and aluminium alloys for protection of steel in marine atmospheres has been of benefit although the application of these types of coatings for protection of steel immersed in seawater is not so usual.

This paper investigates thermal sprayed aluminium coatings in seawater. The following coatings have been tested: arc and flame sprayed Al, Al 5% Mg and 85% Zn 15% Al, both with and without a vinyl sealer.

The corrosion of the coatings was measured by the linear polarisation technique (LPR). Further, the current demands at the anode potential (–1040 mV SCE) were obtained as a function of time.

Polarisation to higher potentials were also made to measure the ability of the coatings to supply current to unprotected steel areas. The results indicate low corrosion rates, e.g. less than 10 μm/year can be obtained at an average sea temperature of 10–15°C, and near stagnant seawater flow conditions.

The current consumption at the anode potential is much lower than obtained on steel. The measured currents indicate a reduction in anode load to 10% of the load necessary for protection of bare steel.

In contrast to the case for immersed painted steel, the damaged part of the total protected area by sprayed Al is supposed to be constant throughout its life time. The Al-coatings are also able to act as sacrificial anodes and protect steel in narrow areas where the current from the anode is hindered.

1. Introduction

Submerged offshore structures have traditionally been protected by a cathodic protection system, which in the Norwegian sector of the North Sea has been of the sacrificial anode type. In order to reduce the amount of anode material, coatings, in particular organic coatings, have begun to be used. Thermally sprayed metal coatings are used to a lesser extent. These coatings have, however, inherent properties which make them particularly interesting in areas where sacrificial anodes cannot be mounted, e.g. on tension legs [1], or in areas where the current from the cathodic protection system may be restricted, e.g. by the geometry.

The following summarise the advantages and disadvantages of thermal sprayed Al and Al-coatings compared with organic coatings:

Advantages:
- Very long lasting protection with very low maintenance cost.
- The coating is hard, strong and is not easily damaged. (Coated parts may be handled by fork lifts.)

- Finished coatings can be used immediately. (No curing time and specific curing temperatures are needed.)
- The coatings resist temperatures up to 400–500°C.
- If properly applied, high adhesion (greater than 12 MPa) may be expected.
- Sprayed Al and Al-alloys have sacrificial abilities on steels in marine environments, which means that if small bare steel areas are exposed the steel will be protected to some extent.
- The spraying process is simple to mechanise.

Disadvantages:
- Before spraying, very accurate surface preparation is required. This gives higher costs than with organic coatings.
- Applying the coating by handheld equipment for long periods is tedious.
- Applying the coating by hand is time-consuming and quite expensive.
- Some rust staining of the coating may occur before the self-sealing process of the coating has taken place.

Thermally sprayed Al-based coatings can also be an alternative to organic coatings on submerged structures in general. In a previous investigation it was shown that an Al-based coating requires a very low cathodic current demand, and a structure covered with 90% Al-based coating would need less than 20% of the amount of anodes needed for a bare steel structure.

In order to rely on a thermally sprayed Al-based coating one has to know its properties under long-term exposure in seawater. Most long term tests reported in the literature have been concerned only with the free corrosion properties of the coatings. Al-based coatings have been found to perform quite well in such exposure tests for up to 19 years [3]. When used in combination with sacrificial anodes or areas of bare steel, the properties of the coating under cathodic as well as anodic polarisation conditions must also be investigated.

The aim of the present study has been to consider the performance of various coatings polarised to a range of potentials from the typical sacrificial anode potential around –1030 mV Ag/AgCl up to a typical steel corrosion potential of –670 mV Ag/AgCl. The coatings have been varied with respect to alloy composition, spraying method, use of organic sealer, surface preparation and the coating thickness. The tests were carried out for up to 18 months in seawater, in order to draw conclusions as to which parameters are of prime importance for a thermally sprayed coating to work satisfactorily in combination with a sacrificial anode system.

2. Experimental

2.1 Preparation of specimens

Two types of specimens were used. A thermally sprayed pipe section of steel was used for the sea test, while thermally sprayed steel panels with dimensions 100 × 75 × 2 mm were used for all the laboratory tests.

The dimensions of the pipe sections are given in Fig. 1. All the specimens were abrasive blasted with crushed iron grit to Sa 3.0 (SIS 055900) before thermal spraying, except for some panels that were blasted to Sa 2 to investigate the influence of surface preparation.

Three types of materials were sprayed as seen in Table 1 together with the spraying method and the nominal thickness of the coatings.

2.2 Environment

In all tests natural seawater was used. The seawater may have seasonal variations especially

Table 1 Types of coatings tested

Abbreviation	Spraying method	Material	Thickness (μm) (Sea test)
A Al	Arc Sprayed	99.6 % Al	150 200
A AlMg	" "	95 % Al, 5 % Mg	50 190
A ZnAl	" "	85 % Zn, 15 % Al	150 -
F Al	Flame sprayed	99.6 % Al	150 160
F AlMg	" "	95 % Al, 5 % Mg	150 160
F ZnAl	" "	85 % Zn, 15 % Al	150 170

in organic activity. The laboratory circulation system water inlet was at 50 m sea depth, and only moderate seasonal temperature variations were observed. The temperature was between 6.5 and 11°C, typically 8°C. The temperature at 20 m depth where the sea test rig was placed varied rather more, and in the summer could reach 14°C.

2.3 Sea test of pipe sections

2.3.1 Potential and current density monitoring

The sea test rig was launched on 15 June 1987 and withdrawn on 4 January 1989. A detailed drawing of the test rig is shown in Fig. 1.

The one bare steel pipe and the thermally sprayed sections were electrically connected to an aluminium anode through a 1Ω resistor. The potential drop across the resistors was measured and the current calculated.

The potentials were monitored with four Ag/AgCl-seawater reference electrodes positioned as shown on Fig. 1.

The currents and the potentials were automatically logged twice a day using a data acquisition system onshore. The coating materials tested in the sea test are listed in Table 2.

A sealer, of vinyl type (Carboline Polyclad 935 Tic coat), was applied by brush onto two of the test sections, as seen in Table 2.

Table 2 Materials tested in the sea test

Coating Materials	Abbreviation	Coating thickness [μm]
1. Abrasive blasted Steel	Steel	0
2. Arcsprayed 99.6%Al	AAl	≈250
3. Arcsprayed AlMg5	AAlMg	≈200
4. Flamesprayed ZnAl15	FZnAl	≈200
5. Flamesprayed 99.6%Al	FAl	≈175
6. Flamesprayed AlMg5	FAlMg	≈175
7. Arcsprayed 99.6%Al sealed	AAlS	≈200
8. Flamesprayed 99.6%Al sealed	FAlS	≈200

2.3.2 Adhesion measurements

The adhesion of the tested coatings were measured by the pull-off method, both before and after the exposure period.

The pull-off method is based on glueing aluminium 'dollies' to the coating. The coating is cut through by a hole saw at the circumference of the dolly, and it is then pulled off and the load recorded. Details of this procedure are described elsewhere [4].

2.4 Laboratory testing of panels in nearly stagnant seawater

Most of the different coatings and influencing parameters were tested in this experiment for a period of 11 months. The panels were placed in parallel positions in a container with circulating natural seawater. The water was exchanged very slowly so that the conditions in the container can best be described as 'nearly stagnant'. A number of specimens were tested at the sacrificial anode potential in the same way as in the sea test. However, in the laboratory tests the coating thickness was also varied between 100 and 250 microns, and some specimens were surface prepared only to Sa 2.0.

Some panel specimen types were tested at higher potentials. The specimen types (see Table 1) were AAl, AAlMg and FZnAl both with and without sealer. The panels were kept at fixed potentials by potentiostats. The chosen potentials were –870, –770 and –670 mV Ag/AgCl. The current was recorded by measuring the potential drop across the connecting resistors, which could be varied from 1 to 1000Ω.

All the six main coatings as listed in Table 1, both with and without a sealer, were tested under free corrosion conditions. The potential was measured with a voltmeter relative to an Ag/AgCl reference electrode. At three different exposure times linear polarisation measurements were carried out to monitor the corrosion rates. The linear polarisation measurements were made with equipment from Tacussel Electronique (type Corrovit) specially designed for such measurements. The polarisation was ± 12.5 mV and the polarisation rate ± 12.5 mV min^{-1}. Polarisation curves were obtained at the end of the test. The potential was started at –1100 mV Ag/AgCl and stepped upwards to –700 mV, in steps of 20 mV and a holding time of 6 min at each step.

2.5 Laboratory testing of panels in slowly flowing seawater

This test was made with a few types of panels to see if slow water flow had any significant influence on the current density at the sacrificial anode potential. The materials tested were AAl, AAlMg and FZnAl without sealer. The test was made in a flow-loop, where the panels were exposed to seawater flowing at a rate of 0.1 ms^{-1}. The panels were connected to an aluminium sacrificial anode, and the current monitored as described above for the panels in nearly stagnant conditions.

3. Results

3.1 Sea test of pipe sections

3.1.1 Electrical current demands

The variation of the potential of the coatings is shown in Fig. 2, and the monitored current densities in Figs. 3 and 4. The dashed lines between the 6th and the 8th month indicate a temporary malfunction of the logging equipment.

The differences in current density between sealed and unsealed specimens, which is the most significant effect found, can be readily seen. The current density of the sealed coatings is lower than 1 mA m^{-2}. No increase of the current with the time for sealed specimens can be noticed, which means that the sealer has not deteriorated within the exposure time.

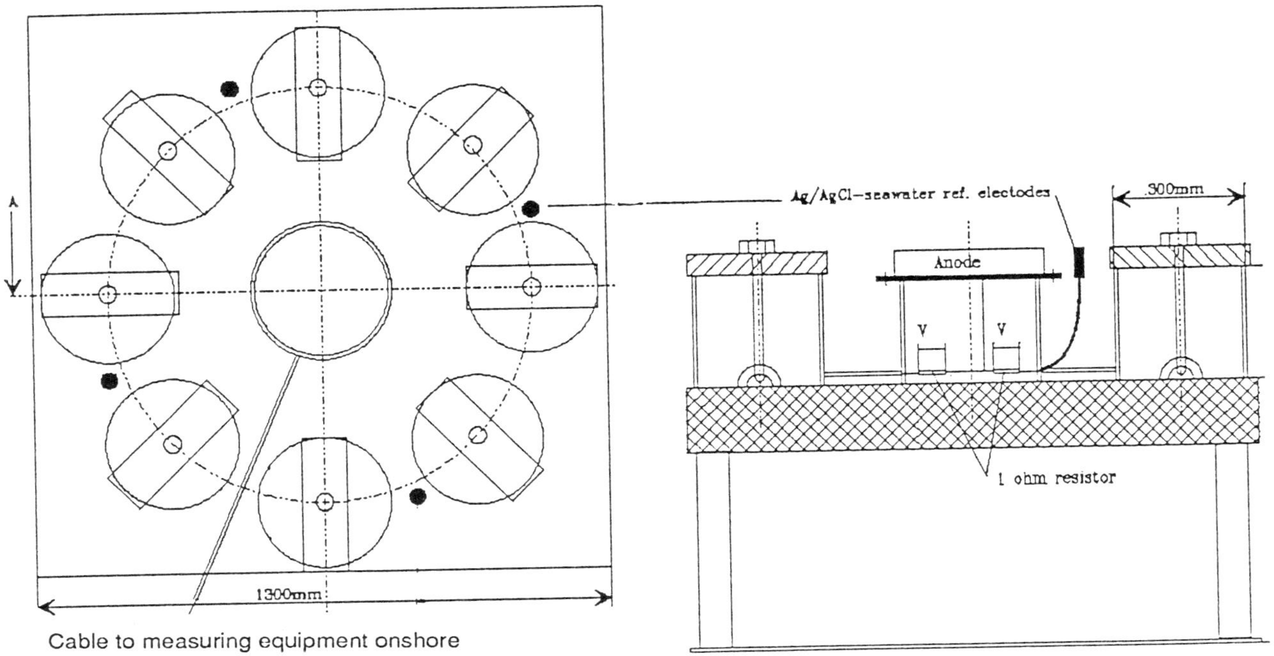

Fig. 1 Sea test arrangement.

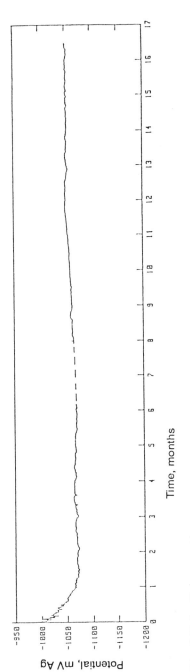

Fig. 2 Potential on coatings.

From Figs. 3 and 4 it can be seen that the current decreases with time and stabilises within an exposure period of 10 months. All the coatings tested required a current density lower than 10 mA m^{-2} at the end of the test period.

3.1.2 Mechanical testing

The adhesion values of the coatings before and after exposure are given in Table 3.

The adhesion values obtained on unexposed samples are considerably lower then the values obtained after exposure. The adhesion on unexposed samples was measured on separate test panels with a thickness of only 2 mm. Some elastic deformation of the test plates during adhesion testing may have occurred, and this could have resulted in high stress intensities at the circumference of the dolly-to-glue interface. This could be the cause of the low adhesion values. The values measured after exposure are quite normal, and indicate no reduction of adhesion during the exposure period except for the values of flame sprayed Al coating with sealer. The specimen with ZnAl-coating was lost during the retrieval operation of the sea test specimens. No adhesion values of exposed ZnAl material have therefore been measured.

3.1.3 Visual observations of sea test specimens

One section of each recovered specimen was cleaned in concentrated HNO_3 to remove the white calcareous scale worms on the coating, especially on the sealed coating specimens. A careful examination revealed that the Al and AlMg coatings showed no sign of deterioration after an exposure period of 18 months in seawater.

3.2 Laboratory testing of panels in nearly stagnant seawater

3.2.1 Cathodic current demands at sacrificial anode potential

The results from this test confirmed the results from the sea test, in that the current density for unsealed specimens decreased with time and eventually decreased to below 10 μA m^{-2} for most of the specimens. For Al and AlMg the sealed specimens were down to below 1 mA m^{-2} but for ZnAl the sealing had no current lowering effect. No systematic effect could be observed of the surface preparation grade.

Table 3 Adhesion values measured on sea test specimens

COATING	Adhesion (MPa)	
	Separate panels not exposed	Exposed section
2 Arc spr. Al	9.0 ± 1.0	15.5 ± 2.3
3 " " AlMg	5.7 ± 0.46	15.2 ± 0.8
4 Flame spr. ZnAl	5.6 ± 1.25	
5 Flame spr. Al	3.5 ± 0.34	7.3 ± 0.65
6 " " AlMg	4.0 ± 0.52	9.25 ± 1.56
7 Arc spr. Al sealed	9.0 ± 2.0	15.4 ± 1.6
8 Flame spr. Al sealed	3.5 ± 0.34	2.8 ± 0.44

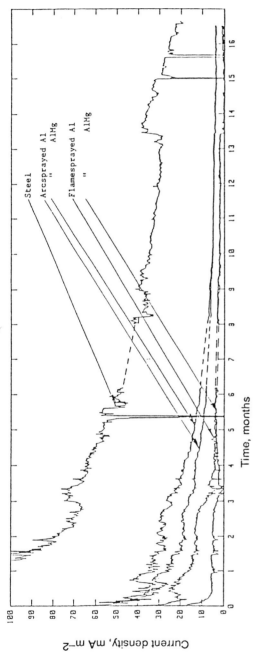

Fig. 3 Cathodic current density of arc and flame sprayed Al coating, arc sprayed AlMg and steel polarised to −1030 mV Ag/AgCl.

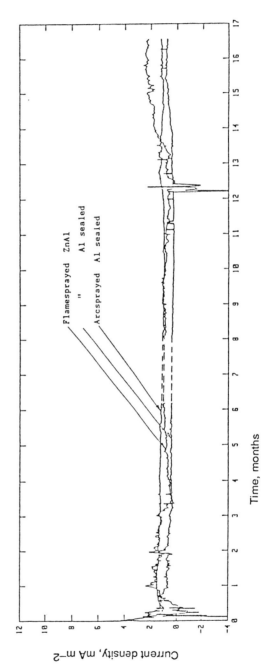

Fig. 4 Cathodic current density of flamesprayed ZnAl and arc and flamesprayed Al with sealer polarised to −1030 mV/AgCl.

3.2.2 Current densities at higher potentials

As outlined in the experimental section, some panels were kept at constant potentials of −870, −770 and −670 mV Ag/AgCl. These potentials set up currents with the time dependence has shown in Figs. 5–7. It should be noted that some of the anodic currents on Al and AlMg increase with time at the beginning of the exposure and decrease with time after having reached a maximum current density. At potentials above −870 mV/AgAgCl, at −770 and −670 Ag/AgCl, severe corrosion is observed, and the decrease in the anodic current density is probably due to exposure of an increasing area of bare steel which is cathodically polarised. The sum of the cathodic and anodic currents will then result in a reduced net current.

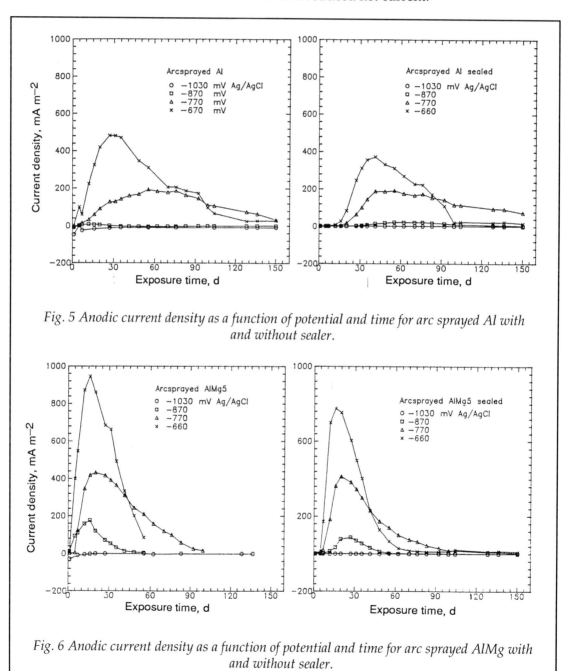

Fig. 5 Anodic current density as a function of potential and time for arc sprayed Al with and without sealer.

Fig. 6 Anodic current density as a function of potential and time for arc sprayed AlMg with and without sealer.

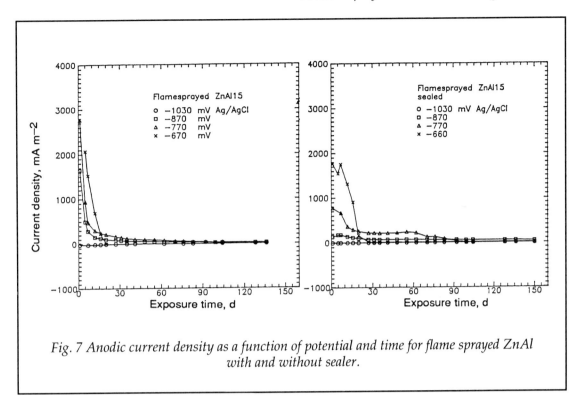

Fig. 7 Anodic current density as a function of potential and time for flame sprayed ZnAl with and without sealer.

In principle, the current densities on ZnAl-coatings seen in Fig. 7 have the same behaviour as the current on Al and AlMg, but the increase of the current density to a maximum happens very quickly and only the decreasing part of the current density curves is seen. The decrease is due to exposure of bare steel—as discussed for Al and AlMg-coatings.

The maximum current densities seen in Figs. 5–7 are those available when the coating is used as a 'sacrificial anode'.

3.2.3 Corrosion potential vs time
The free corrosion potential as a function of time is shown in Figs. 8–10. The potential varies to some extent with the test material, and the time before a relatively stable potential is reached is quite long. The potentials of the tested materials after 11 months are given in Table 4.

3.2.4 Corrosion rates at free corrosion
The corrosion rates of the materials measured by linear polarisation are presented in Figs. 11 and 12. The corrosion rates were also calculated based on extrapolation of polarisation curves. These rates are compared in Table 5 with the rates calculated from linear polarisation measurements.

The agreement between the two methods is quite good, and point to the conclusion that the corrosion rates for all coatings after 11 months is ca. 5µm/year or less. In the first months the corrosion rates are somewhat higher, particularly for arc sprayed ZnAl.

3.3 Effect of water flow
Water velocity was observed to influence the polarising current at the anode potential (21038 mV Ag/AgCl). This can be seen from Table 6 where the current density at nearly stagnant conditions and at ≈ 0.1 ms^{-1} water flow, are compared. The lowest values are for a water flow of 0.1 ms^{-1}.

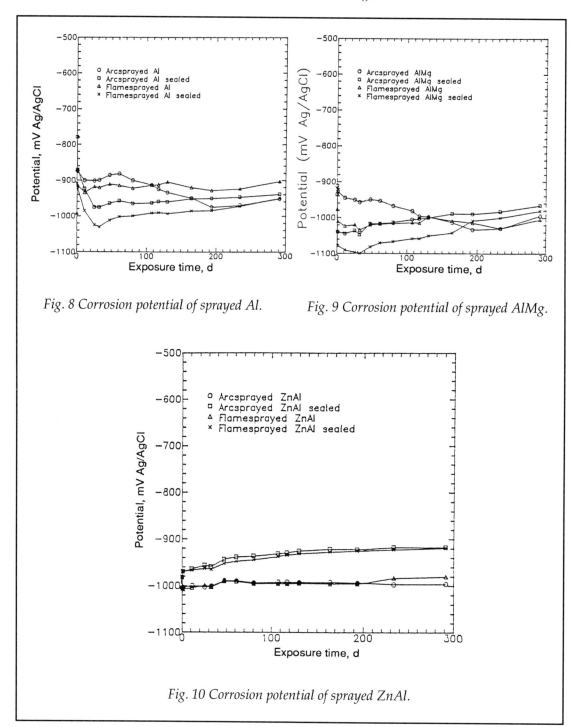

Fig. 8 Corrosion potential of sprayed Al.

Fig. 9 Corrosion potential of sprayed AlMg.

Fig. 10 Corrosion potential of sprayed ZnAl.

3.3.1 Visual observations of laboratory test specimens

Cathodically polarized samples (−1030 mV Ag/AgCl) and samples at free corrosion with Al or AlMg coating were found to be in excellent condition at the end of the test. No deterioration or blistering was observed; only a few white spots indicating that the corrosion has not come to a complete stop. The ZnAl-covered panels showed severe blistering.

At −870 mV Ag/AgCl the Al and the AlMg covered panels are still in a very good shape,

Table 4 Corrosion potentials after 11 month exposure

Coating material	Abbreviation	Potential mVAg/AgCl
Arcsprayed Al	AAl	-950
" Al sealed	AAlS	-940
Flamesprayed Al	FAl	-910
" Al sealed	FAlS	-950
Arcsprayed AlMg	AAlMg	-995
" AlMg sealed	AAlMgS	-970
Flamesprayed AlMg	FAlMg	-1000
" AlMg sealed	FAlMgS	-1010
Arcsprayed ZnAl	AZnAl	-995
" ZnAl sealed	AZnAlS	-920
Flamesprayed ZnAl	FZnAl	-980
" ZnAl sealed	FZnAlS	-920

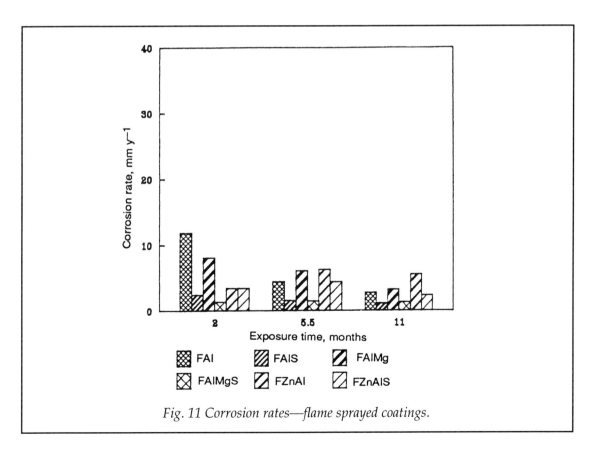

Fig. 11 Corrosion rates—flame sprayed coatings.

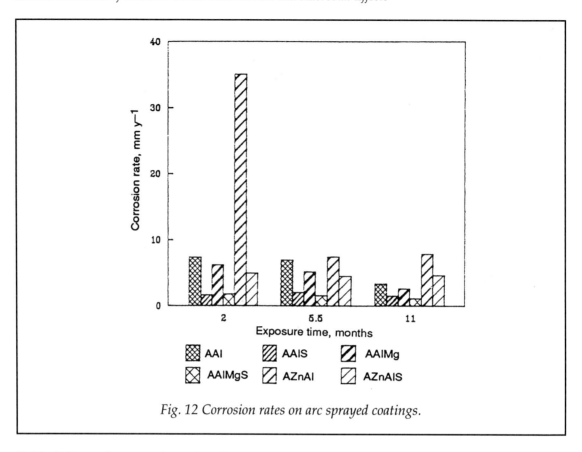

Fig. 12 Corrosion rates on arc sprayed coatings.

Table 5 Corrosion rates based on linear polarisation (LPR) and on polarisation curves. Exposure time ≈ 11 months

Coating	LPRmeas. µm/year	Polarization µm/year
AAl	3.3	3.35
AAlS	1.5	1.25
AAlMg	2.6	2.2
AAlMgS	1.1	0.95
AZnAl	7.8	4.9
AZnAlS	4.7	1.15
FAl	2.7	2.0
FAlS	1.1	1.0
FAlMg	3.2	2.25
FAlMgS	1.3	1.7
FZnAl	5.5	3.85
FZnAlS	2.3	1.1

Table 6 Current density (mA m^{-2}) at ≈ –1030 mV Ag/AgCl after 7 months exposure

Coating	Water flow	
	Nearly stagn.	0.1 m/s
AAl	8	0.60
AAlMg	1	0.19
FZnAl	7.5	5.05

but there are more corrosion products on AlMg than on Al. ZnAl has corroded quite heavily.

At –770 mV Ag/AgCl also Al and AlMg panels show heavy corrosion and corrosion products. When these panels were cleaned from corrosion products it was observed that spots of bare steel were visible on the surface.

At –670 mV Ag/AgCl all panels rusted heavily, indicating a near complete loss of the coating during the test period.

4. Discussion

4.1 General

Thermally sprayed coatings to be used on submerged parts of offshore structures must be selected according to a number of properties. The most important of these properties are:

- the corrosion potential;
- the cathodic current demand at low potentials;
- the corrosion rate at low potentials;
- the corrosion rate at high potentials;
- the adhesion strength.

In addition, other factors such as cost, speed of application, quality variations under practical application conditions, etc. must be considered. These factors will not be discussed here.

4.1.1 The corrosion potential

The corrosion potential of a metal coating must be quite low and reasonably stable. It is not necessary that the potential is as low as for sacrificial anodes, but it should be lower than the protection potential for the steel substrate, i.e. lower than –800 mV Ag/AgCl.

4.1.2 The cathodic current demand at low potentials

When a thermally sprayed coating is used in combination with a cathodic protection system, the current drained by the coating will be an important parameter. The lower the current demand the better, as it reduces the number of sacrificial anodes needed for a full lifetime protection.

4.1.3 The corrosion rate at low potentials

The corrosion rate will determine the lifetime of a coating with a given thickness. It must be sufficiently low to provide the lifetimes of 30 years or more, that have recently been specified as design lifetimes offshore. As shown in the previous section, the corrosion rate can be determined from polarisation measurements at the corrosion potential or above. Under cathodic protection the corrosion rate cannot be measured electrochemically. Weight loss measurement is an alternative, but due to the weight increase associated with changes in the oxide, the method is not accurate. In the present study no weight loss measurements have been done. The corrosion rates under cathodic protection are said to be smaller than the values measured at the corrosion potential.

One should also be aware that for Al-based coatings under cathodic protection, in contrast to steel, there is no well-defined protection potential below which the corrosion rate is zero. It has been shown for solid aluminium [5] that the corrosion rate is reduced when cathodic protection is applied, but it never goes to zero—it only reaches a minimum.

4.1.4 The corrosion rate at high potentials

At potentials above the free corrosion potential of the coating the corrosion rate will increase, and this is easily determined from measurements of the net anodic current. In short time periods one must expect the potential to stay above the free corrosion potential of the coating. This can happen immediately after the structure is submerged. Minor areas of bare steel will need some time before an efficient layer of calcareous deposit is formed and the potential drops below the protection potential of the steel. Mechanical damages to the coating exposing new areas of bare steel at a later date can also lift the potential to a higher level for a short period. Under such conditions it is important that the corrosion rate of the coating is also low at high potentials. Even if the residence time at these high potentials is short (of the order of a few weeks), the corrosion damage to the coating can be unacceptable if the corrosion rate is several hundreds of mA m^{-2}.

4.1.5 The adhesion strength

It is obvious that this is one of the most important properties of a coating. If the coating adheres poorly to the substrate, it is not to be recommended, even if the electrochemical properties are favourable. In this study the adhesion strength has been measured on some of the samples, but blistering observed during visual inspection of the samples after a lengthy time exposure is a strong indicator of poor adhesion.

4.2 Effects of coating alloy composition

Of the three alloys tested, Al, AlMg and ZnAl, with the composition as given in Table 1, the Al coating seems to have the overall best performance. All the alloys are acceptable regarding the corrosion potential. ZnAl has the lowest cathodic current demand, due to the corrosion potential being very close to the sacrificial anode potential, but the current demands for Al and AlMg decrease to below 10 mA m^{-2} after a few months in the sea test, which is also acceptable. The laboratory test shows somewhat lower values for AlMg compared to Al, and the reason for this is not known.

The corrosion rates at the free corrosion potential for Al and AlMg are of the order 5µm/y after 2 months, decreasing towards 2–3µm/year after 11 months. This is quite acceptable, as it indicates lifetimes of more than 30 years for a 150µm thick coating. ZnAl has somewhat higher corrosion rates, especially during the first few months.

At higher potentials ZnAl is observed to corrode extremely quickly. At –870 mV Ag/AgCl the corrosion rate is already more than 1000 mA m^{-2}, which is not acceptable, AlMg is somewhat better, but this alloy also showed some pitting at –870 mV, with an initial current density rising to nearly 200 mA m^{-2} at maximum. Al is even better, showing very low corrosion rates at –870 mV. At –770 mV and –670 mV all materials corrode, but Al has the lowest rates.

Blistering was observed on ZnAl under free corrosion conditions, which was not unexpected from previous observations with the same material [6]. However, panels of ZnAl held at the sacrificial anode potential also showed blisters, indicating that weak cathodic polarisation does not improve the adhesion properties of this coating. Neither Al nor AlMg showed blistering at any potential. At –770 and –670 mV vs Ag/AgCl all coatings deteriorated, mainly as a result of heavy corrosion.

In conclusion, ZnAl is not to be recommended, because of its poor adhesion and high corrosion rates near and above the free corrosion potential. Al and AlMg are both acceptable, but Al has the lowest corrosion rates.

4.3 Effects of the spraying method

The spraying method does not seem to have a strong influence on the overall performance of the coatings. No systematic effects were observed regarding the corrosion potential or the corrosion rates. In the sea test, the cathodic current demand is somewhat lower for the flame sprayed coatings than for the arc-sprayed coatings throughout the whole test period, but in the laboratory test this was not so obvious. The measured adhesion strengths are clearly in favour of the arc-sprayed coatings, with values about twice as high as the 6.9 MPa quality specification limit used for flame-sprayed Al coatings on the Hutton tension leg platform, according to Fischer et al.[7].

4.4 Effects of sealer

A vinyl type sealer is found to have a clear positive effect on the coating performance. This is most evident in the data for the cathodic current demand and the corrosion rates at the free corrosion potential.

The sealed coatings of Al or AlMg have a cathodic current demand below 1 mA m^{-2} throughout the whole test period, while the unsealed coatings drain current densities of 20–30 mA m^{-2} over the first 4–6 months, before the natural sealing effect reduces these values towards 10 mA m^{-2}, or lower. For an offshore structure with an Al-based coating on some parts of the surface and bare steel or organic coating on the rest, it is a great advantage that the Al-based coating drains a very low current density in the beginning. This will reduce the number of sacrificial anodes needed to protect the whole structure within a given time.

The corrosion rates at the free corrosion potential for sealed coatings are typically 30–50% of the values without the sealer after 11 months of exposure. This shows that a sealer, despite its low thickness, may be effective over a longer period of time than expected, as long as the potential stays at the free corrosion potential or lower.

At higher potentials the sealer has almost no effect in reducing the corrosion of the coatings.

4.5 Effects of surface preparation

Effects of surface preparation were studied only in the laboratory test. On Al the samples prepared with sandblasting to Sa 2.0 instead of Sa 3.0 showed somewhat higher cathodic current density demands, but this effect was less obvious on AlMg.

What is more important is that none of the badly prepared Al or AlMg samples showed any sign of coating deterioration after exposure times of about one year.

4.6 Effect of the coating thickness

The only effect of the coating thickness noticed was a somewhat higher cathodic current density demand for thick coatings on flame sprayed Al. On arc-sprayed Al-coatings and on the AlMg coatings no clear effect was observed.

4.7 Effect of the flow rate

The strong reduction in the cathodic current demand upon increasing the flow rate from nearly stagnant to 0.1 ms^{-1} is possibly due to a lowering of the corrosion potential of the coatings. Fischer et al. [7] performed a number of tests at 0.1 ms^{-1} flow rate and observed corrosion potentials much lower than in the present tests at stagnant conditions. Even if a lower cathodic current demand is advantageous, we cannot draw the conclusion that the coatings will perform better at increasing flow rates. We have no data on the corrosion rates at 0.1 ms^{-1} and we do not know the behaviour at higher flow rates.

5. Conclusions

Flame and arc sprayed coatings of 99.6 % Al, AlMg with 5% Mg and ZnAl with 15% Al have been tested for up to 18 months in seawater. The coatings have been tested with and without sealer, and the thickness has been varied between 100 and 200μm. In the test period the cathodic current demand at sacrificial anode potential has been monitored, as well as the corrosion potential and corrosion rates for freely corroding samples. The corrosion rates have also been monitored at potentials up to –670 mV Ag/AgCl.

The results have been discussed from the point of view that the coatings are candidate barrier coatings for submerged offshore structures in combination with a sacrificial anode cathodic protection system. The main conclusions are:

1. All the coatings show decreasing cathodic current demand with the time at –1030 mV Ag/AgCl. After 18 months the current density is about 5 mA m^{-2} for unsealed coatings and less than 1 mA m^{-2} for sealed coatings.

2. The corrosion potentials vary somewhat with time, but stay, mainly between –1000 and –900 mV Ag/AgCl for all coatings, with the highest values for the Al coatings.

3. The corrosion rates at the free corrosion potential are reduced with time from about 10μm to about 2–3μm/y for the Al and AlMg coatings. For ZnAl the corrosion rate remains about three times greater.

4. At anodic polarisation to 2870 mV Ag/AgCl, the Al coatings still have a very low corrosion rate, while AlMg show a temporary large increase in the corrosion rate, up to 200 mA m^{-2} of anodic current output. At –770 mV, and in particularly at –670 mV Ag/AgCl, both Al and AlMg corrode quite rapidly. At the upper potential the lifetime is of the order of a few weeks. ZnAl coatings corrode heavily at all these potentials.

5. Blisters were observed on ZnAl under cathodic protection and in free corrosion conditions. The Al and AlMg coatings showed no blistering or deterioration at any potential except for the corrosion damage at –770 mV and above.

6. The adhesion values of the coatings after 18 months in the sea were highly acceptable except for the sealed flame sprayed Al sample. The arc-sprayed samples had readings at about 15 MPa, and were somewhat superior to the flame-sprayed samples in general.

7. Both Al and AlMg coatings are acceptable as barrier coatings in combination with a sacrificial anode system. Al is probably the best choice due to the lowest corrosion rate at weak anodic polarisation. ZnAl is not recommended, due to blistering at low potentials and extremely high corrosion rates at higher potentials.

8. Arc sprayed coatings have better adhesion than the flame sprayed coatings. Other properties are little influenced by the spraying method.

9. The use of a vinyl sealer greatly improves the barrier properties of the coatings. The cathodic current is reduced by an order of magnitude, and the corrosion rate at the free corrosion potential is lowered by a factor of 2–3. The sealer has little effect at higher potentials.

10. The coating thickness and the surface preparation variations had only minor influences on the properties of the coatings.

6. Acknowledgement

This work has been supported by Statoil and by Elf Aquitaine Norge A/S. The authors wish to thank the management of these companies for their permission to publish this paper.

References

1. W. H. Thomason, Cathodic protection of submerged steel with thermal-sprayed aluminium coatings, Materials Performance, 1985, **24**, 3.
2. P. O. Gartland, Cathodic protection of aluminium-coated steel in seawater, Materials Performance, 1987, **26**, 6.
3. 'Corrosion Tests of Flame-Sprayed Coated Steel', 1974, 19 Year Report — AWS C2.14-74, American Welding Society, Miami, FL.
4. METCO, 1969, Flame spray handbook, 1969, Vol. 1, p. 46-50, METCO Inc., Westbury, Long Island, NY.
5. B. Sandberg and A. Bairamov, Cathodic protection of aluminium structures. Report from the Swedish Corrosion Institute, 1985-04-01.
6. B. A. Shaw and P. J. Moran, Characterization of the corrosion behavior of zinc–aluminium thermal spray coatings, Materials Performance, 1985, **20**, 22.
7. K. P. Fischer, W. H. Thomason and J. E. Finnegan, Electrochemical Performance of Flame Sprayed Aluminium Coatings in Seawater. Corrosion NACE '87, San Francisco, paper 360.